Cambridge Elements

Elements of Paleontology
edited by
Brenda R. Hunda
Cincinnati Museum Center

WHAT DOES GRAPTOLITE ORIGINATION AND EXTINCTION REVEAL ABOUT THE CAUSE OF THE LATE ORDOVICIAN MASS EXTINCTION?

Charles E. Mitchell
University at Buffalo

H. David Sheets
Merrimack College

Michael J. Melchin
St. Francis Xavier University

Chris Holmden
University of Saskatchewan

CAMBRIDGE
UNIVERSITY PRESS

Shaftesbury Road, Cambridge CB2 8EA, United Kingdom

One Liberty Plaza, 20th Floor, New York, NY 10006, USA

477 Williamstown Road, Port Melbourne, VIC 3207, Australia

314–321, 3rd Floor, Plot 3, Splendor Forum, Jasola District Centre, New Delhi – 110025, India

103 Penang Road, #05–06/07, Visioncrest Commercial, Singapore 238467

Cambridge University Press is part of Cambridge University Press & Assessment, a department of the University of Cambridge.

We share the University's mission to contribute to society through the pursuit of education, learning and research at the highest international levels of excellence.

www.cambridge.org
Information on this title: www.cambridge.org/9781009684101
DOI: 10.1017/9781009684064

© Charles E. Mitchell, H. David Sheets, Michael J. Melchin, and Chris Holmden 2025

This publication is in copyright. Subject to statutory exceptionand to the provisions of relevant collective licensing agreements, with the exception of the Creative Commons version the link for which is provided below, no reproduction of any part may take place without the writtenpermission of Cambridge University Press & Assessment.

An online version of this work is published at doi.org/10.1017/9781009684064 under a Creative Commons Open Access license CC-BY-NC 4.0 which permits re-use, distribution and reproduction in any medium for non-commercial purposes providing appropriate credit to the original work is given and any changes made are indicated. To view a copy of this license visit https://creativecommons.org/licenses/by-nc/4.0

When citing this work, please include a reference to the DOI 10.1017/9781009684064

First published 2025

A catalogue record for this publication is available from the British Library

ISBN 978-1-009-68410-1 Hardback
ISBN 978-1-009-68405-7 Paperback
ISSN 2517-780X (online)
ISSN 2517-7796 (print)

Additional resources for this publication at www.cambridge.org/mitchell-et-al

Cambridge University Press & Assessment has no responsibility for the persistence or accuracy of URLs for external or third-party internet websites referred to in this publication and does not guarantee that any content on such websites is, or will remain, accurate or appropriate.

For EU product safety concerns, contact us at Calle de José Abascal, 56, 1°, 28003 Madrid, Spain, or email eugpsr@cambridge.org

What Does Graptolite Origination and Extinction Reveal about the Cause of the Late Ordovician Mass Extinction?

Elements of Paleontology

DOI: 10.1017/9781009684064
First published online: December 2025

Charles E. Mitchell
University at Buffalo

H. David Sheets
Merrimack College

Michael J. Melchin
St. Francis Xavier University

Chris Holmden
University of Saskatchewan

Author for correspondence: Charles E. Mitchell, cem@buffalo.edu

Abstract: This Element assesses the macroevolutionary turnover of paleotropical planktic graptolites during the Late Ordovician Mass Extinction (LOME) via automated sequencing and capture-mark-recapture modeling. Graptolites exhibited a succession of turnover pulses (sensu Elizabeth Vrba) that were coincident with the main phases of the Hirnantian glaciation and during which the Diplograptina experienced declining metapopulation size, elevated extinction, zero species originations, and, ultimately, complete extermination. Concurrently, the Neograptina (latest Katian temperate zone immigrants) exhibit pulses of both extinction and adaptive radiation. Thus, the LOME involved intense species selection and the wholesale alteration of the clade diversity structure of a major element of the zooplankton. The LOME likely was not a direct effect of ocean anoxia or sampling bias but rather resulted from Hirnantian climate change, which altered nutrient supplies and plankton community compositions, resulting in ecological displacement and loss of habitat that together drove the succession of turnover pulses. This title is also available as open access on Cambridge Core.

Keywords: evolutionary rates, turnover pulse, metapopulation decline, species selection, macroevolution

© Charles E. Mitchell, H. David Sheets, Michael J. Melchin, and Chris Holmden 2025

ISBNs: 9781009684101 (HB), 9781009684057 (PB), 9781009684064 (OC)
ISSNs: 2517-780X (online), 2517-7796 (print)

Contents

1 Introduction	1
2 Methods	11
3 Results	22
4 Discussion	41
5 Conclusions	60
6 Afterthought: The Late Ordovician Timescale	61
References	63

An Online Appendix for this Element is available at
www.cambridge.org/mitchell-et-al

1 Introduction

Recent geochemical studies suggest that large environmental changes took place during the Late Ordovician in the interval leading up to and during the Late Ordovician mass extinction (LOME). These changes include a secular decline in global temperatures, most likely in response to reduced atmospheric pCO_2, possibly related to ultramafic weathering or enhanced carbon burial (possibly due in part to land plant and marine green algal biomass expansion; see the recent review by Algeo & Shen Jun 2023) or some combination of those factors, which culminated in continental-scale glaciation. In addition, some recent work suggests that Late Ordovician cooling may have been interrupted by warming episodes created by the catastrophic outpouring of CO_2 and H_2S from massive volcanicity, although this remains controversial (e.g., Bond & Grasby 2017, Dahl *et al.* 2021, Zhou Yu-ping *et al.* 2024). The Late Ordovician also apparently experienced large changes in sea level as well as changes in nutrient cycling, carbon burial, and phytoplankton community composition, among other possible effects. These oceanographic and nutrient cycling effects are manifest in pronounced facies changes, stable isotopic excursions, including the widely documented *Hirnantian isotopic carbon excursion* (HICE), as well as reported perturbations in nitrogen, sulfur, uranium, and others (e.g., Melchin *et al.* 2013, Harper 2023, Young *et al.* 2023).

Testing alternative hypotheses about the direct cause (or causes) of the LOME requires a clear and precise understanding of the timing, rate, and selectivity of the extinctions. As the current, regionally heterogeneous response to global warming demonstrates, however, no single section or region can provide an adequate proxy for global change. Thus, a full assessment of mass extinction dynamics must include a global analysis that integrates the individual paths of biotic response to the local facies and habitat manifestations of global environmental change, while also compensating as much as possible for the artifacts that arise from sequence stratigraphic effects on the preserved record (Holland & Patzkowsky 2015, Holland 2016, Zimmt *et al.* 2021, Holland 2023). Yet many recent papers regarding the LOME have nonetheless extrapolated from small regional studies or even single sections to global-scale conclusions about Late Ordovician climate and mass extinctions while taking little or no account of local-scale basin dynamics or potential sequence stratigraphic effects on the record. It is not surprising, then, that the literature is now replete with conflicting explanations for this momentous episode in Earth history.

The specific goals of the present research are, first, to use the paleotropical synthesis of graptolite species occurrence data employed by Melchin *et al.* (2017) to assess the rate, timing, and selectivity of graptolite species turnover within a global, high-resolution, sample-based composite timescale; and secondly, to apply that sample-based composite timescale to make a precise set of comparisons between the resulting global history and the patterns of environmental change in local sections across the paleotropics and thereby test the principal alternative hypotheses about the drivers of mass extinction during the Late Ordovician.

1.1 Background

From their analysis of the Late Ordovician record, Brenchley et al. (1994, 2001, 2003) suggested that the Late Ordovician mass extinction comprised two pulses, and this interpretation has been widely (if somewhat imprecisely) adopted in studies of the LOME (e.g. Sheehan 2001, Finnegan *et al.* 2012b, Melchin *et al.* 2013, Harper *et al.* 2014, Luo Gen-ming *et al.* 2016, Zou Cai-neng *et al.* 2018a, Bond & Grasby 2020, Hu Dong-ping *et al.* 2020, Chen Yan *et al.* 2021, Kozik *et al.* 2022a, Harper 2023, Hu Rui-ning *et al.* 2024, among many others). Not coincidentally, the Late Ordovician glacial interval, although part of a long-term and gradually increasingly severe glacial epoch, is now widely regarded as having culminated in two major glacial advance cycles within the Hirnantian itself separated by a brief warm period that took place early in the interval of the *Metabolograptus persculptus* Biozone (reviewed in Ghienne *et al.* 2014; see Fig. 1). Nonetheless, Chen Xu *et al.* (2005b), based on species occurrence data from a set of four sections in South China, found that graptolite extinction in this interval was dominated by a single early Hirnantian pulse. Likewise, Wang Guang-xu *et al.* (2019) concluded from their global review of Late Ordovician faunas that turnover among brachiopods during the LOME was dominated by a pulsed, although somewhat diachronous, extinction around the beginning of the Hirnantian that was then followed by an extended period of turnover as the newly evolved faunas tracked the changing Hirnantian and early Rhuddanian environments. These results reinforce the concerns, highlighted by Holland and Patzkowsky (2015), that we should have about the degree to which the "two pulse" model is a real feature of the LOME rather than an artifact generated by the effects of glaciation on the rock record itself.

1.1.1 Graptolites and the Late Ordovician Mass Extinction

Although meaningful analyses of faunal dynamics have been conducted using genera, it is obvious that species-level data are preferable when the record is sufficiently complete. The graptolite fossil record is such a case (see, for example, Sadler *et al.* 2011). Based on their analysis of a global compilation of graptolite species ranges, Foote *et al.* (2019, p. 1049) estimated that for the

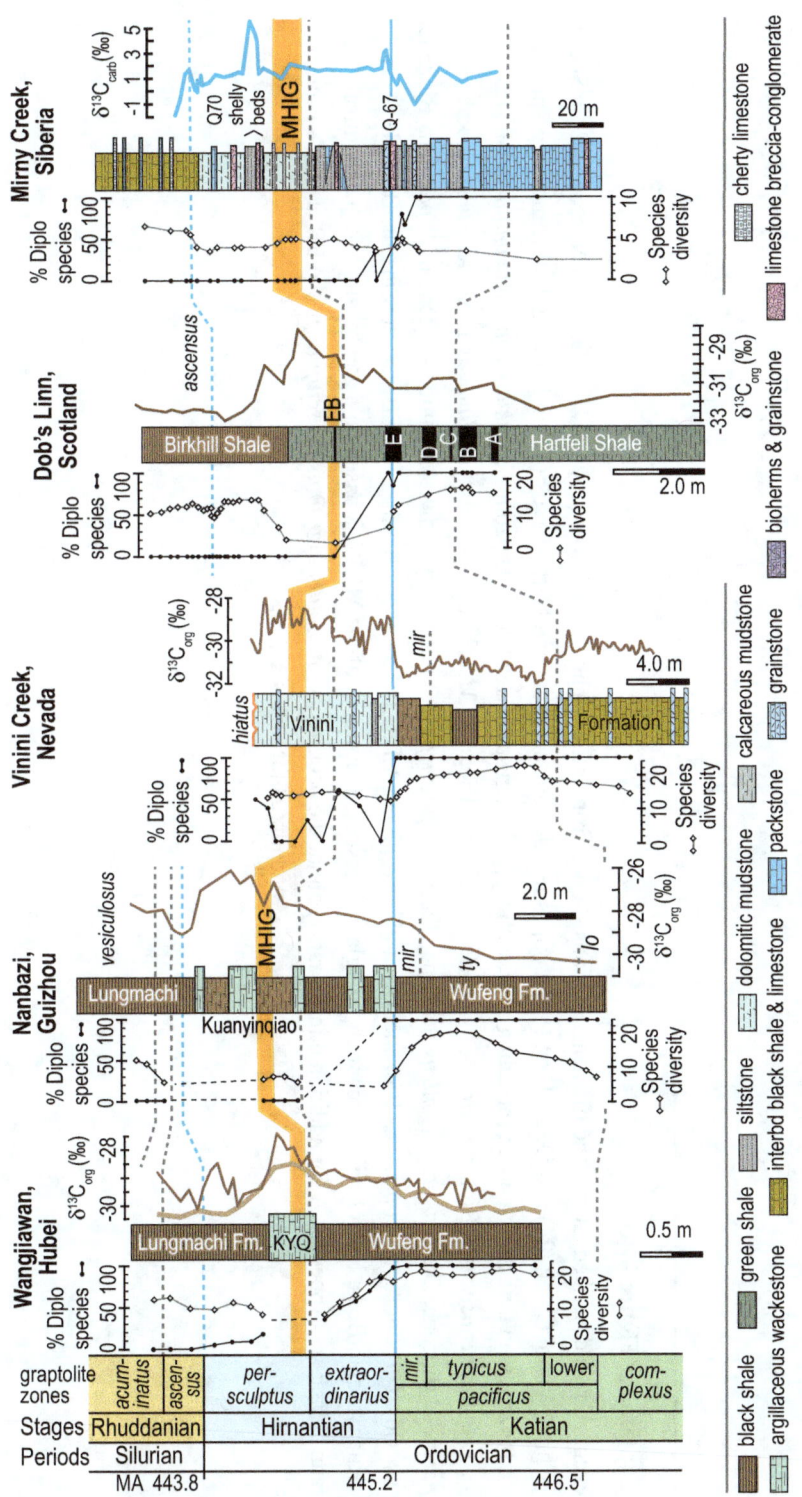

Figure 1 Observed trajectories in graptolite species diversity and turnover in clade composition relative to lithostratigraphy and the Hirnantian carbon isotopic excursion (HICE) at five intensively studied sections from across the paleotropics (see Fig. 2 for site locations). Sections exhibit distinctly different patterns of graptolite faunal turnover relative to lithological change and the trajectory of the C-isotopic

Caption for Figure 1 (cont.)

excursion, as described later in the Element. Sections shown, from left to right: Wangjiawan (25) and Honghuayuan/Nanbazi (12) in South China; Vinini Creek (24) and Dob's Linn (6) on opposite sides of Laurentia; and Mirny Creek (19) in Kolyma. Full names of graptolite biozones (in descending order, including abbreviations): *Cystograptus vesiculosus* Biozone, *Parakidograptus acuminatus* Biozone, *Akidograptus ascensus* Biozone, *Metabolograptus persculptus* Biozone, *Metabolograptus extraordinarius* Biozone, *Paraorthograptus pacificus* Biozone, *Diceratograptus mirus* Subzone, *Tangyagraptus typicus* Subzone (ty); lower unnamed subzone of *P. pacificus* Biozone (lo), *Dicellograptus complexus* Biozone. Other abbreviations: A,B,C,D,E, Anceps Bands A–E; EB, Extraordinarius Band; KYQ, Kuanyinqiao Beds; MHIG, mid-Hirnantian interglacial episode. Placement of the MHIG is based on a combination of geochemical and faunal data; see Appendix A for additional information about the occurrence of *Metabolograptus persculptus* in the EB at Dob's Linn and Appendix B for data sources.

species that dwelt within the region of preserved Ordovician and Silurian strata, approximately "75% of Ordovician–Silurian graptoloid species have been sampled and that, of those known from more than one resolvable stratigraphic level, c. 85% of their original durations are represented by their composite stratigraphic ranges." Foote *et al.* concluded that relative to other studied animal groups, graptolites have one of the most completely sampled fossil records.

Additionally, as macrozooplankton, graptolites offer a unique perspective on environmental dynamics within the surface regions of the world's oceans. Individual species are known to have been widely distributed across the paleotropics (Chen Xu *et al.* 2003, Boyle *et al.* 2017) (Fig. 2) and, as such, their fates were likely to have been intimately intertwined with similarly global features of the Late Ordovician environment (Cooper *et al.* 2012, 2014, Crampton *et al.* 2016, Crampton *et al.* 2018). Observed graptolite species diversity in the paleotropics declined dramatically through the LOME from a peak of about 80 species known in mid Katian time (~ 446 Ma) to a Hirnantian nadir of about 20 species, barely 1.5 Myr later (Sadler *et al.* 2011). The composition of the faunas changed dramatically as well. Late Ordovician graptolite faunas were populated by species from two major clades: the Diplograptina and the Neograptina, which had diverged during the Darriwilian, approximately 15 Myr prior to the LOME events (Mitchell *et al.* 2007a, 2009, Štorch *et al.* 2011, Maletz 2023). In the late Katian, prior to the mass extinction, the Diplograptina were diverse and common

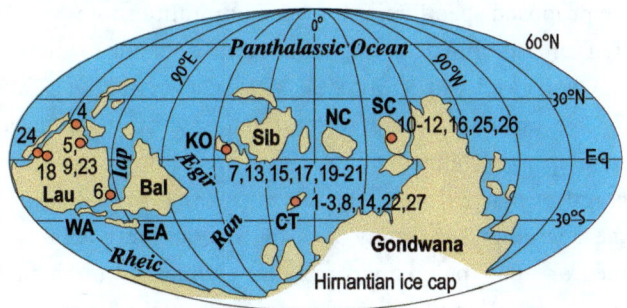

Figure 2 Location map of 27 Late Ordovician to early Silurian graptolite-bearing sections studied here. Paleoplates from which data were used in the present study are labeled with the following abbreviations: Bal, Baltica; CT, Chu Ili-Tien Shan terrane; EA, East Avalon terrane; Iap, Iapetus Ocean; KO, Kolyma-Omolon terrane, Lau, Laurentia; NC, North China; SC, South China; WA, West Avalon terrane. Sites numbered as in Melchin *et al.* (2017); see Appendix 2 for data sources and sample-by-sample species occurrence data from these sites.

throughout the paleotropics, where they comprised some 16 genera and were the only planktic graptolites present (Chen Xu *et al.* 2003, Goldman *et al.* 2011, Melchin *et al.* 2011, Goldman *et al.* 2014). The Late Ordovician Neograptina, in contrast, have a sparse fossil record and are only known from about ten species within two genera that were entirely confined to mid-high paleolatitude regions. However, coincident with the onset of the LOME in the latest Katian (i.e., within the *Diceratograptus mirus* Subzone of the *P. pacificus* Biozone), the Neograptina invaded the paleotropics and rapidly became the dominant graptolites (Chen Xu *et al.* 2005a, Goldman *et al.* 2014, Sheets *et al.* 2016). By the end of the Hirnantian, the Diplograptina were entirely extinct (Melchin & Mitchell 1991, Melchin 1998, Chen Xu *et al.* 2006b, Finney *et al.* 2007, Goldman *et al.* 2011, Bapst *et al.* 2012).

1.1.2 Distinguishing the Local and the Global

In general it appears that the macroevolutionary rates and clade turnover in graptolite faunas exhibit long-term rhythms that are comparable in duration with global Milankovitch-driven climate cycles (Crampton *et al.* 2018). Graptolite species turnover rates shifted to higher values during the LOME, in synchrony with the HICE and Hirnantian glaciation, and those elevated rates then persisted for the remainder of the clade's history (Cooper *et al.* 2014, Crampton *et al.* 2016). Those results clearly implicate climate oscillations as a major driving force in graptolite extinction and several prior studies have proposed specific ties between graptolite ecology and the likely effects of Hirnantian climate change that could account for their extinction during the LOME (Melchin & Mitchell 1991, Chen Xu *et al.* 2005b, Finney *et al.* 2007). Nevertheless, the precise global pattern of timing and rate of turnover in graptolite faunas relative to Late Ordovician environmental change has not been well characterized at a temporal resolution sufficient to assess the degree of synchrony with the Hirnantian glacial advance and retreat cycles or other paleoenvironmental changes through that interval. Furthermore, regional and local-scale environmental effects (including but not limited to those associated with preserved facies shifts or hiatuses) have overprinted the record of global climate change. For instance, detailed local records of graptolite species turnover relative to the HICE and the mid-Hirnantian interglacial episode (which may provide approximate, independent measures of synchroneity among sections – see Holmden *et al.* 2013, Melchin *et al.* 2013, Ghienne *et al.* 2014) reveal disparate patterns of change in species diversity and faunal composition. These patterns may be grouped broadly into three sets (Figs. 1, 2):

- Abrupt loss of all graptolites coincident with early Hirnantian facies change; examples include sections at Blackstone River, Laurentia [site 4], Ojsu Spring, Kazakhstan [22] and Ludiping, South China [16].
- Graptolites remain common in the Hirnantian strata but faunas undergo a rapid loss of diversity and abrupt replacement of diplograptines by neograptines through the Katian-Hirnantian transition interval (ex: Vinini Creek, Laurentia [24], Mirny Creek, Siberia [19], Durben well, Kazakhstan [8], Honghuayan, South China [12]).
- Faunas show an extended period of species diversity decline and clade replacement over the course of the latest Katian to mid-Hirnantian – namely through the entire first Hirnantian glacial megacycle (ex: Wangjiawan, South China [36], Dob's Linn, Laurentia [6]).

In all of these cases diversity decline commences in the mid to upper part of the uppermost Katian *Paraorthograptus pacificus* Biozone; roughly in step with the onset of the HICE, however, the disappearance of species is earlier and more rapid in the more on-shore sites and latest in the more oceanic sites (see Sheets *et al.* 2016, figs. S4, S5). Nonetheless, species losses are present in the deep sites that do not appear to be purely a result of facies and habitat displacement (e.g., at Vinini Creek, Sheets *et al.* 2016), and at several of these sites species that appeared to go extinct in this initial pulse reappear as Lazarus taxa in the *Metabolograptus persculptus* Biozone in association with the falling limb of the HICE in the upper Hirnantian (Štorch *et al.* 2011). Thus, many of the apparent early Hirnantian losses may be artifacts of sampling or habitat displacement or both (Mitchell *et al.* 2007b) rather than true extinctions, consistent with sequence stratigraphic considerations (Holland & Patzkowsky 2015, Holland 2016, 2023).

This ambiguity about the timing of extinctions is amplified by the individual failures of even these well-studied sections. Two examples will suffice here. The Mirny Creek section (the most complete and most fossiliferous of the seven Omulev Uplift sections studied in detail by Koren *et al.* 1983) is orders of magnitude thicker than the other principal graptolite-bearing Late Ordovician sections, and graptolites are rare in the Mirny Creek strata (Fig. 1). Most species are represented in each of the Omulev region collections by one to five specimens (Koren *et al.* 1983), despite a cumulative collection effort amounting to 20 person-months (T. Koren' personal communication to CEM, 2005). Thus, it is likely that only the most common species were recovered, and that if some diplograptine species persisted into the Hirnantian there as rare individuals (see Koren' & Sobolevskaya 2008), as we know they did elsewhere, these relict species almost certainly would not have been recovered.

The late Katian and early Hirnantian record at the Dob's Linn section, which contains the global stratotype for the base of the Silurian System and has figured prominently in discussions of the LOME, is even more highly incomplete. Graptolite faunas of the late Katian *P. pacificus* Biozone, although reasonably diverse (Fig. 1), are recoverable from only two relatively thin black shale intervals (Anceps Band C and D; Williams 1982). This zone has an estimated duration of some 1.8–3.38 My in the GTS 2012 and 2020 timescales, respectively (Cooper & Sadler 2012, Goldman et al. 2020). The base of the Hirnantian is not recorded precisely, but most likely lies somewhere between Band D and Band E, which contains the first record of *Metabolograptus extraordinarius*, and thus probably corresponds to a level within the early Hirnantian *M. extraordinarius* Biozone (Williams 1982, Melchin et al. 2003, Chen Xu et al. 2006a). Restudy of the very small fauna preserved within the overlying, ~ 2 cm-thick Extraordinarius Band indicates that it contains *Metabolograptus persculptus* and is thus of late Hirnantian, *M. persculptus* Biozone age (Melchin et al. 2003; see also Appendix A of this Element for further details). Accordingly, Band E is the only level from which early Hirnantian graptolites may be recovered at Dob's Linn. It is thick enough to be subsampled (Fig. 1) but nonetheless represents only a small portion of the total duration of this biozone. Furthermore, sedimentological evidence suggests that Band E was deposited during an interglacial episode (Armstrong & Coe 1997), and is thus atypical of the glacial conditions that prevailed during the early Hirnantian. Thus, the critical late Katian to mid Hirnantian LOME interval is represented at this site by five samples, none of which appear to sample the first main glacial advance. Since resolving the order of a set of events requires at least as many samples as events, the record at Dob's Linn is grossly insufficient to resolve the extinction history of the 17 species present in the *P. pacificus* interval there. It is surprising, therefore, to see graptolite diversity data from this section cited as evidence that a spike in Hg abundance at this site indicated that the mass extinction was caused by an abrupt pulse in volcanic activity (Bond & Grasby 2020). Such a causal relationship simply cannot be determined from those data.

From the foregoing it is clear that no single section, not even the best, can serve adequately as a proxy for the full environmental and evolutionary history that we seek to understand. Accordingly, we have compiled a global composite based upon the full record of graptolite species sightings, occurrence by occurrence, through 27 sections from across the paleotropics (Melchin et al. 2017) and herein report on the pattern of graptolite turnover extracted from that composite. As we discuss more fully later, this high-resolution composite overcomes some of the failures of the record imposed by local sections and permits a close, and precisely timescaled comparison between geochemical environmental proxies from those sections and the composite macroevolutionary history of graptolites.

1.1.3 Sample-Based Graptolite Occurrence Dataset

The dataset that we employed for this study of graptolite species dynamics through the LOME is the same dataset employed in Melchin *et al.* (2017). It comprises 3,508 presence-absence records for 105 species from 633 collections distributed among 27 measured sections. The selected sections are those that exhibit relatively continuous sedimentation through the LOME interval, are rich in graptolites, and have been documented via detailed systematic study of those faunas. We have compiled this dataset from the published record of species occurrences; in several cases, however, we have augmented that record by additional collections from the sections (especially at Dob's Linn) and, for most of these collections, by direct restudy of the original graptolite material. We have applied a uniform taxonomy to the species occurrences insofar as this is possible based on the published record and our ability to restudy those collections. The study sites occur within the region ~ 20° north and south of the paleoequator and encompass approximately 210° in paleolongitude. Data from higher-latitude regions are available from very few sites and include too few species in common with the paleotropical faunas to be reliably integrated by the automated sequencing approach that we employed to form a unified composite of species occurrences (see the "Methods" section later in the Element). Thus, although conceptually of value in the present context, these occurrences are not treated quantitatively herein.

The number of species that cooccur within individual samples varies considerably among the samples employed for this study and most samples contain few species (Fig. 3), some of which are confined to single sections. Such species are not stratigraphically informative and the age of the samples in which they occur is constrained only by the presence of other, more widely shared species – either in those samples or in other samples in the same section. In the present study we employ the occurrences of stratigraphically informative species to form an ordered sequence of samples by the methods described later in the Element. The ranked frequency distribution of informative species diversity per sample follows the expected negative log-normal distribution for sampled species diversity more generally, with an R^2 of 0.9627 between informative species diversity and the log of the number of samples. The median per-sample diversity of stratigraphically informative species is 4.0 (Fig. 3A). Similarly, histograms of species sighting frequency (Fig. 3B) indicate that most stratigraphically informative species occur in few samples (Fig. 3B), consistent with

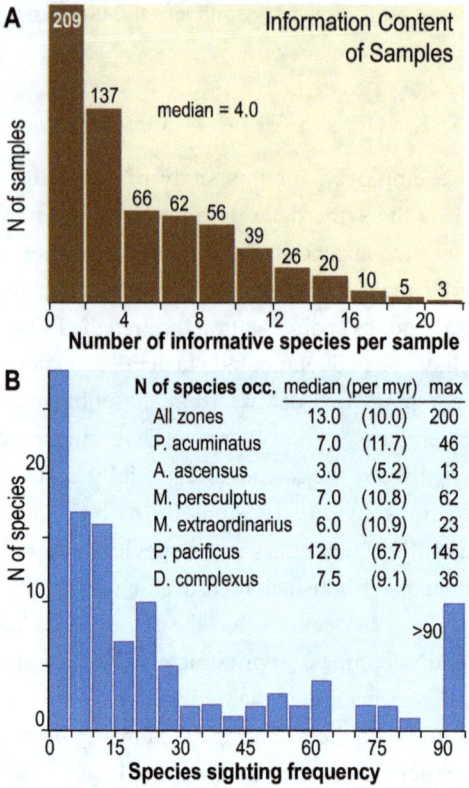

Figure 3 A Histogram of the number of stratigraphically informative species per sample (ordinate). Thirty-five samples contain only species that are not shared with samples from other sections (i.e., species that are unique to single localities) and thus have zero informative species. The position in the ordinal composite of such samples is constrained only by samples above or below them that do contain stratigraphically informative species. The ordinal position of 67 percent of samples is constrained by the joint occurrence in those samples of three or more species and that of slightly less than half is constrained by five or more. **B**, Histogram of the sighting frequency (i.e., the number of recorded occurrences) for each species through its full range within the dataset, together with tabulated median, median/myr, and maximum sighting frequencies within individual biozone intervals and the study interval as a whole; for example, *Appendispinograptus supernus* (which is the most widely reported species in the set) is reported in 200 samples within the full dataset and in 145 samples within the *P. pacificus* Biozone.

the findings of Foote *et al.* (2019); however, Fig. 3B also shows a second mode of species that are moderately to widely occurring with 50 or more occurrences in the sampled sections. The median number of occurrences per species ranges from three to 12 among biozones in the latest Ordovician and earliest Silurian, which, although somewhat variable, shows no obvious trend through the study interval, particularly when treated as sightings per million years (Fig. 3B; we describe the sources of the zone durations in Section 2.2). Furthermore, the number of sightings per million years for individual species during the Hirnantian are similar to the overall median for the dataset. Accordingly, the available data appears to be sufficient to warrant the analyses described in the following sections.

2 Methods

2.1 Determining Species Ranges

We determined graptolite species ranges by use of an automated sequencing approach called Horizon Annealing (Sheets *et al.* 2012, Melchin *et al.* 2017). Horizon Annealing (HA) is a modification of the more widely used constrained optimization (CONOP) approach developed by P. M. Sadler and colleagues (Kemple *et al.* 1995, Sadler & Kemple 1995, Sadler *et al.* 2003). These automated sequencing approaches are based on the contention that the first and most fundamental task for measuring species durations in the fossil record is to determine the order of those events – the species' global first and last appearance events (FAD and LAD, respectively). That set of ordered events is then scaled relative to a convenient timescale – generally one measured in millions of years and derived from geochronologically dated samples that can also be placed within the ordinal sequence of events (see the discussion in Goldman *et al.* 2020 and references cited therein).

Detailed descriptions of the HA ordination process are given in Sheets *et al.* (2012) and Melchin *et al.* (2017), but briefly, both the HA and CONOP automated sequencing techniques seek to determine the global ordinal sequence of a set of species' FADs and LADs in such a way that the ordination minimizes the sum of all the species ranges while also accounting for the observed overlaps among the species ranges. CONOP does this by explicitly ordering species' FADs and LADs based on the order of these events within individual sections. HA, in contrast, orders all available samples (not just range ends) based on the species contents of the samples (constrained by stratigraphic position within sections) and the implications of those occurrences for the overlap of species ranges, seeking to minimize the number of range overlaps while again accounting for all observed within-sample co-occurrences and range overlaps. The ordinal position of 67 percent of samples is constrained by the joint occurrence in those samples of three or more species and that of slightly less than half is constrained by five or more species (Fig. 3A).

For the purpose of sample ordination in HA, we may define the term "taxa" broadly to include not only species occurrences but also any other observable event that appears to have regional or global chronostratigraphic significance, such as volcanic ash layers, unique lithostratigraphic markers (such as sequence boundaries or other distinctive event beds), or chemostratigraphic events, among other possibilities. In the process of constructing the ordination employed here, Melchin *et al.* (2017) included the Kuanyinchiao Bed, which is a lithostratigraphically distinctive unit that is widely distributed in the mid-Hirnantian succession of South China (Chen Xu *et al.* 2004, Chen Qing *et al.* 2014) as well as the rising and falling limbs of the HICE, each treated as discrete taxa (see Melchin *et al.* 2017 for further discussion).

For this analysis all of the samples and taxa (whether graptolite species or other event types) were weighted equally. This approach did not privilege 'key' or zonal index taxa and did not assume that the FAD of *M. extraordinarius* (or of any taxon) is everywhere the same age. Thus, the ordination employed here is an unbiased maximization of all the available information in the dataset about the sequence of graptolite FADs and LADs and does not rely upon any particular assumption about the correspondence of species appearances or disappearances to Late Ordovician glacial or other events (such as the HICE) or to the timing of those events. Each is allowed to find its own level, as it is best fit by the data overall. The resulting ordination integrates species occurrences across regional facies gradients and paleoplates within a sequential framework that maximizes the fit of all the available data to the modeled species durations. It is this relaxation of the assumption of synchroneity among species range ends and the integration of occurrences across regional facies gradients (from on-shore sites to outer shelf and slope sites) that maximizes the likelihood of overcoming taxon range truncations imposed by eustatic changes in sea level and the resulting incompleteness of the geological record (Holland 2020, Zimmt *et al.* 2021).

The Horizon Annealing composite employed here has three additional features that are noteworthy in the present analyses. Firstly, the HA composite is amenable to jackknife analysis, in which sections are removed from the dataset, one section at a time, and the ordination solution is then recalculated. The standard deviation of the variation in the placement of each sample across the set of jackknife replicates may serve as a measure of uncertainty in the position of that sample in the ordination (Melchin *et al.* 2017). This uncertainty naturally differs among samples based on the faunal content of the sample and its stratigraphic context. In the present case, the median standard deviation in horizon placements (omitting the lowest and highest 16 percent of ordinal positions, where variance in position appears to be constrained by edge effects) is 8.14 ordinals. This jackknife uncertainty value provides a useful gauge of the reliable resolution of the ordination and thus served as a guide to selecting a minimum temporal bin duration for our turnover analysis.

Secondly, because the analyzed sample set retains all occurrences of each of the graptolite taxa in the studied sections, rather than only their FADs and LADs, the final ordination includes information about the frequency with which taxa occur through the succession and among sites. That information, then, can be used to assess the likelihood of encountering an extant taxon in an observational interval. This is a key piece of information needed to refine estimates of species origination and extinction rates and to assess their interdependency with changes in completeness of the dataset through the glacial interval (Foote 2001). We will return to this topic later in our discussion of Capture-Mark-Recapture analysis.

Thirdly, because HA orders samples rather than taxa, and because all samples from a given section must remain in their observed stratigraphic order, HA can carry any sample through the ordination process, irrespective of whether that sample contains species or other properties that are informative about their position in the ordinal sequence. Thus, the HA composite can include occurrences of species that are unique to single samples or sections in so far as these horizons are constrained by other horizons in the same sections. More importantly, this property of HA allows ordinal placement of samples that bear locally unique chemostratigraphic, paleoenvironmental, or lithostratigraphic information of value to the interpretation of the geological and evolutionary history of the species of concern. We make use of this feature in our discussion of the possible environmental causes of the observed graptolite species turnover. Taken together these features provide a well-grounded evidentiary basis from which to assess species turnover dynamics and facilitate discrimination between local and global patterns of turnover and their relationships to paleoenvironmental change.

2.2 Temporal Scaling of the Ordinal Composite

In the HA composite, each sample occupies a unique position in the ordered sequence of samples. In order to provide species ranges in millions of years, we must convert that ordinal sequence to a timescaled succession. We estimated median durations (in Myr) for *D. complexus* through *C. vesiculosus* biozones based on those given in GTS2012 (Cooper & Sadler 2012, Melchin et al. 2012) and combined those estimates with additional estimates of zone duration. These latter include estimates derived from geochronological age dates (Ling Ming-xing *et al.* 2019, Du Xue-bin *et al.* 2020) and astrochronological interval durations (Lu Yang-bo *et al.* 2019, Jin Si-ding *et al.* 2020, Zhong Yang-yang *et al.* 2020). The 3.38 Myr duration of the *P. pacificus* Biozone in the GTS2020 timescale (Goldman *et al.* 2020) is unusually long compared to that in the GTS2012 and to other estimates of the duration of this zone. Consequently, we did not use the GTS2020 zone durations in our timescaling process although we did adopt 443.07 Ma as the current best estimate of the age of the beginning of the Silurian. After determining the median duration of each zone, we placed the position of zone boundaries in the composite based on the ordinal position of the FAD of the key index taxon for each zone. Working from the GTS2020 estimated age of the beginning of the Silurian (i.e., the start of the *A. ascensus* Biozone; 443.07 Ma), we used the median durations to provide estimates of the geochronological age of each of the other zonal boundaries. Finally, we used linear interpolation to convert the ordinal positions of all horizons within each of the biozones to a corresponding geochronological age. The resulting intervals in the scaled composite that encompass the set of samples from each of the graptolite biozones or subzones are chronostratigraphic units – that is, they are our best estimate of the particular intervals of geological time during which the strata of the biozones were deposited. Following the standard terminology of the International Commission on Stratigraphy, we refer to these intervals in our scaled composite as chrons and subchrons and label them by the trivial epithet of the eponymous species of the corresponding biozone (e.g., "Extraordinarius Chron," and so on).

We divided the resulting temporally scaled composite into twenty-four analytical intervals (referred to herein as bins), each with a duration of 210,000 years.[1] These temporal bins contain an average of 24 horizons and a range of 10 and 61 horizons per bin. Thus, all bins are longer than the jackknife uncertainty in the ordinal position of horizons and most are several times longer. Most of the biostratigraphic units recognized through the latest Ordovician and early Silurian formed during intervals on the order of 0.6 Myr and are represented in the dataset by two or three bins. The relatively long *P. pacificus* Chron (with its three subchrons), which has a median estimated duration of 1.8 Myr in the datasets we employed, is represented by eight or nine bins. The temporal resolution of our chosen binning is comparable to that which Chen Xu *et al.* (2005b) employed in their analysis of graptolite species turnover in the Yangtze Platform region of China and similar to the duration of the moving window that Copper *et al.* (2014) and Crampton *et al.* (2016, 2018) employed in their analyses of graptolite species diversity and turnover.

---

[1] See note about late-arriving updates to the Late Ordovician timescale, located following the Conclusions section.

Recognizing that uncertainty in the ordinal location of events in the Melchin *et al.* (2017) HA composite is substantial relative to the resolution in our binning scheme, we constructed two alternative placements of the 24 equal-duration bins relative to the composite sequence: binning schemes, B1 and B2 (Fig. 4). We also employed a third, substantially different binning scheme, B3, that we describe further later. These alternatives provide an indication of the effects of bin boundary placement on the estimated rates and timing of species turnover. In the first binning scheme, B1, we placed the bin boundaries such that the first bin spans the beginning of the Complexus Chron and the start of the 14th bin coincides with the global FAD of *Metabolograptus extraordinarius* (i.e., to the start of the Extraordinarius Chron). That level in the composite is equivalent, within error, to the position of the Hirnantian GSSP at Wangjiawan, South China, which also lies within B1 bin-14. Thus, bin B1-14 represents the beginning of the Hirnantian Age as formally defined (see Melchin *et al.* 2017). In B2, we offset the bins upward by approximately one-half a bin so that the start of the first bin coincides with the FAD of *D. complexus* (= the beginning of the Complexus Chron) and the FAD of *M. extraordinarius* falls within B2-13. This bin placement retains the Hirnantian GSSP level within B2-14, and thus, does not impose synchroneity between the global first appearance of *M. extraordinarius* and the start of the Hirnantian Age.

The offset between the B1 and B2 schemes results in differences in the total number of bins that incorporate Hirnantian samples (six in B1 and seven in B2) and changes the extent to which the 24 bins extend into the early Silurian (Fig. 4). They also shift slightly the overlap between the Mirus Subchron and our bins. This is noteworthy because previous work on the LOME indicates that both the turnover events and related environmental changes that are associated with the LOME (including disruptions of the carbon system reflected in the HICE) accelerated during the Mirus Subchron. That upsurge in the pace of LOME-related change will thus appear in slightly different places among the three binning schemes; it occupies the 12th and 13th bins in B1, only the 12th bin in B2, and the 9th–11th bins in B3 (more on this last scheme later).

Foote (2000) demonstrated that short-term changes in sampling intensity or probabilities of taxon recovery affect estimates of rates of origination and extinction. As we noted earlier, the number of samples included in each bin differs among bins and it also differs among binning schemes for similar points in the time series (Fig. 4). This arises because the biozones that we used to scale the composite themselves differ in duration and because the number of samples within each biozone differs substantially among biozones independently of their duration (Table 1). Consequently, the temporal spacing of the ordinal horizons differs through the time series. We have used two different approaches to assess the

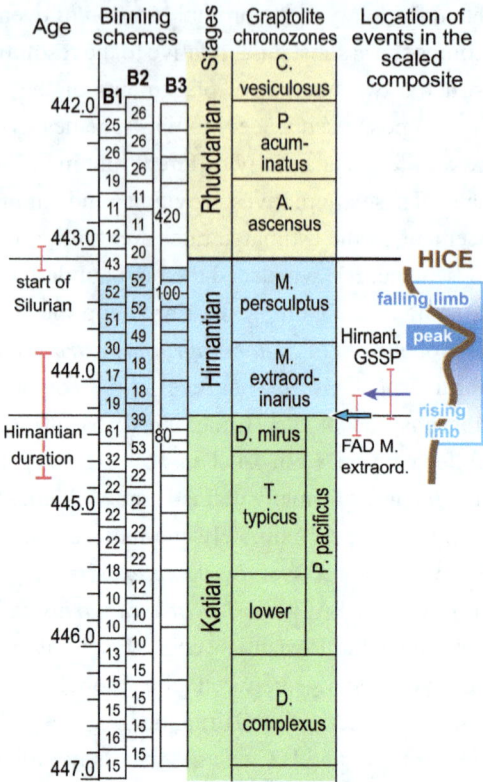

Figure 4 Timescale employed for the analysis of graptolite species diversity and turnover during the late Katian and Hirnantian ages (Late Ordovician Epoch) to early Rhuddanian Age (Llandovery Epoch, Silurian Period). Geochronological ages (in Ma) and chronozone durations employed herein based on the GTS2020 timescale (Goldman et al. 2020) and data from the literature (see Table 1 and text for discussion). Uncertainty error bars (±2σ) shown for the geochronological age of the beginning of the Silurian, duration of the Hirnantian (based on calculations herein), the first appearance datum (FAD) of *Metabolograptus extraordinarius* and the Hirnantian GSSP in the timescale, along with the location of the rising limb, peak, and falling limb of the widespread Hirnantian d^{13}C isotopic excursion (HICE), all based on the ordinal position of global events, including the FADs and LADs of the rising and falling limbs of the HICE, in the Melchin et al. (2017; see also Appendix B) composite. Also shown is the temporal alignment of the three sets of 24 analytical bins (B1-B3) employed for the diversity analysis. All bins in B1 and B2 are 210 Kyr in duration but encompass different numbers of horizons (N from 10–52); those of

B3 are variable in duration (80–420 Kyr) but each includes 24 horizons. Hirnantian bins are shaded and the beginning of the first Hirnantian-aged bin in impact of the differing levels of sampling among bins. First, by shifting bin boundaries to form the B1–B2 sets we not only altered the placement of environmental events relative to those boundaries (concentrating or spreading episodes), we also shifted the density of data within bins (Fig. 4) relative to those events. For instance, the four bins that span the beginning of the Hirnantian in B1 contain 32, 61,19 and 17 horizons, whereas those in B2 contain 22, 53, 39, and 18. We also constructed the B3 binning, in which each of the 24 bins contains exactly 24 horizons. Those bins also average 210 Kyr, as in the other two bin sets, but range in duration from 80 to 420 Kyr. This clearly shifts the density of data in a dramatically different way, equalizing sampling per bin at the cost of equal duration bins. Features of the turnover history of graptolites shared among these three different bin sets must necessarily be largely independent of related sampling effects, at least at the ~ 210 Kyr resolution of the dataset. An obvious alternative is to analyze the data via a method, such as capture-mark-recapture analysis, that directly addresses uneven rates of taxon recovery, which we describe in the following section.

2.3 Estimation of Species Turnover Dynamics and Capture-Mark-Recapture Analysis

We employed two related approaches to quantifying species diversity, origination, and extinction rates. The first is the set of metrics (referred to as "face-value metrics" in the following text) advocated by Foote (2000). These metrics are simple to calculate from presence-absence data and provide a relatively unbiased basis for the comparison of macroevolutionary rates and taxonomic diversity based on the fossil record. In this scheme, the chosen occurrence

---

Caption for Figure 4 (cont.)

the B1 set (B1-14) is aligned with the beginning of *M. extraordinarius* Chron, whereas the beginning of the first bin in B2 and B3 is aligned with the beginning of the *D. complexus* Chron. Based on the placement of samples marking the LAD of the rising limb and FAD of the falling limb of the HICE (see Melchin et al., 2017 for discussion of the coding of the segments of the HICE) the late HICE peak displayed at Wangjiawan (site 25) and Dob's Linn (site 6) and shown in Fig. 1, occupies an interval from about 443.73 ± 0.19 Ma, coeval with the start of the Persculptus Chron, to about 443.62 ± 0.10 Ma at the beginning of the falling limb of the HICE in mid Persculptus Chron time. At other sites, such as Vinini Creek (site 24) and Blackstone River (site 4), the broader HICE peak commences near the beginning of the Hirnantian (LaPorte et al., 2009; see also Fig. 1 in this Element).

Table 1 Biozones in the late Katian to early Rhuddanian (Rhud.) stages of the Late Ordovician and early Silurian systems and biozone durations employed for timescaling the Horizon Annealing composite. Modeled ages are based on median zone durations and the GTS2020 age estimate for the beginning of the Silurian Period.

			Biozone	Ordinal HA composite			Zone duration estimates (Myr)						Time-scaled HA composite		
				HA score at base	ordinal position at base	N of horizons in zone	Cooper et al. 2012	Du et al. 2020	Ling et al. 2020	Lu et al. 2019	Jin et al. 2020	Zhong et al. 2020	Median duration (my)	Duration per ordinal (ky)	Modeled age at base (Ma)
Silurian	Rhud.		Cystograptus vesiculosus	0.9637	610	–	–	–	–	–	–	–	–	–	441.90
			Parakidograptus acuminatus	0.8499	538	72	0.93	–	–	–	0.26	–	0.60	0.826	442.49
			Akidograptus ascensus	0.7994	506	32	0.43	–	–	–	0.73	–	0.58	1.813	**443.07**
Ordovician	Hirnantian		Metabolograptus persculptus	0.5434	344	162	0.6	–	–	0.52	0.72	0.84	0.66	0.407	443.73
			Metabolograptus extraordinarius	0.4708	298	46	0.73	–	0.20	1.22	0.55	0.39	0.55	1.196	444.28
	Katian	Paraorthograptus pacificus Zone	Diceratograptus mirus Subzone	0.3460	219	79	–	–	0.27	–	–	0.27	0.27	0.342	444.55
			Tangyagraptus typicus Subzone	0.1943	123	96	–	–	–	–	–	0.94	0.94	0.976	445.49
			Lower Subzone	0.1501	95	28	–	–	–	–	–	0.59	0.59	2.107	446.08
			P. pacificus (undiv.)	0.1501	95	203	1.86	3.7	1.03	1.8	–	1.80	1.80	0.887	446.08
			Dicellograptus complexus	0.0553	35	60	0.6	–	–	0.82	–	1.27	0.82	1.367	446.90

record is taken at face-value (hence the descriptor) and taxa are censused in a bin-by-bin scheme according to which of the bins' boundaries they cross, as follows. Of the number of taxa that occur within a bin, some are bottom crossers (i.e., are known from older bins) and are symbolized as N_b; of those some will go extinct within the bin (= N_{bL}, for **b**ottom crosser and **L**ast occurrence) and the remainder will pass into younger bins (cross the top boundary, =N_{bt}). Some of these latter may only be known from bins above and below but are treated as range-through taxa and included in the quantity N_{bt}. Additionally, some taxa within a bin cross (or are inferred to have crossed) the top of the bin (N_t), and these include species that first appeared (**F**) within that bin (N_{Ft}). Taxa that occur only within one bin ('singletons') cross neither bin boundary (N_{FL}). Their number strongly depends upon bin duration and sampling completeness and thus increase the vulnerability of taxonomic diversity and evolutionary rate metrics to sampling artifacts (among other problematic effects; see Foote, 2000). Accordingly, we follow Foote's practice and exclude singletons from our diversity and evolutionary rate calculations. Given these notations, estimated mean standing diversity is calculated as $(N_{bL}+N_{Ft}+2N_{bt})/2$ and estimated per capita origination ($\hat{p}$) and extinction ($\hat{q}$) rates (respectively) are $-\ln(N_{bt}/N_t)/\Delta T$ and $-\ln(N_{bt}/N_b)/\Delta T$, where ΔT is the bin duration. For the sake of convenience, we plot the estimated diversity, origination, and extinction rate values at bin midpoints.

The Diplograptina went extinct in the late Hirnantian and we are unable to report an extinction rate for the bin in which the last known species occurs just as one cannot calculate an extinction rate for the final bin in a set of bins: it requires information that does not exist. Similarly, the Neograptina invaded the paleotropics in the latest Katian (Goldman et al. 2011) and we therefore are unable to report an origination rate at that appearance or to distinguish in situ origination from immigration during this early segment of Neograptine resurgence in the paleotropics. Thus, the $\hat{p}$ and $\hat{q}$ time series reported from the face-value and CMR analyses for these clades are one bin shorter at both ends than the time series of species diversity with which they are associated in our figures and do not fully capture the timing of their initial immigration or final extirpation due to the logical limitations of binning and the calculations involved.

We calculated mean standing species diversity and per capita rates separately for each of the Late Ordovician to early Silurian graptolite clades (the Diplograptina and the Neograptina) following the systematic treatment of these graptolites given by Štorch et al. (2011) and Melchin et al. (2011) as well as for the full set of planktic graptolite species present in this interval. In order to provide additional phylogenetic context for the LOME events, we also counted the number of species appearances and disappearances in each bin for the three constituent subclades of the Diplograptina (the Dicranograptoidea, Diplograptoidea, and Climacograptoidea). We also separated the Neograptina into the clade Retiolitoidea and its subtending stem group, which we refer to as 'stem-group neograptines,' namely the Normalograptidae including species of the genus *Normalograptus*, and all other taxa that root below the common ancestor of the Retiolitoidea sensu Melchin *et al.* (2011). The Retiolitoidea is a diverse clade that in the LOME interval was represented overwhelmingly by *M. extraordinarius* and related species of the paraphyletic Neodiplograptidae (*Metabolograptus, Neodiplograptus, Korenograptus,* and *Paraclimacograptus*).

We have calculated the face-value metrics described earlier for each of the three different binning schemes as a rough gauge of the turnover history of graptolites through the LOME and also as a means to assess the role of binning (and its effect on sampling) on the calculated time series. These face-value metrics have a number of shortcomings, however. Most significantly for our present purposes, they do not incorporate information about the probability of sampling taxa through the study interval (Foote 2003). If sighting probabilities are very unequal over time or space, or differ among taxa, then the face-value metrics will not capture the macroevolutionary dynamics accurately (Foote 2000, 2001, 2003). For instance, the omission of sighting information from the analyses forces one to assume that observed first and last occurrences represent the actual time of origin or extinction of the taxa. Considering the global sea level fall and related facies changes that are known to have occurred during the Late Ordovician (summarized in the introduction), it is likely that sighting probability was variable through this interval and that this incompleteness may have distorted the record of species losses through the Late Ordovician (Finnegan *et al.* 2012b, Holland & Patzkowsky 2015, Zimmt *et al.* 2021). We consider this effect on the record of graptolite turnover during the LOME in greater detail in the discussion section of the paper.

Capture-Mark-Recapture (CMR) analysis provides a model-based approach that offers a more nuanced reading of the original presence-absence data (reviewed in Liow & Nichols 2010). CMR was developed for wildlife population assessment and adapts readily to the study of species-level biodiversity. It makes use of maximum likelihood to simultaneously estimate likelihoods of species sighting (p_i) as well as parameters known as survival and seniority. Survival (φ_i) is the probability that a taxon extant in the i-th interval is still extant in the following interval (i+1). Seniority (γ_i) is the probability that a taxon extant in the *i*-th interval was also extant in the preceding interval (*i*–1). Per-interval, per-taxon probabilities of extinction and origination may be derived from these parameters: the probability of extinction is $1-\varphi$, and that of origination is calculated as $1/(1+\gamma_{i+1})$ –1 (Connolly & Miller 2001). To facilitate their comparison with the estimated per capita origination ($\hat{p}$) and extinction ($\hat{q}$) rates described earlier, we divide (or rescale) the per-interval probabilities obtained from our CMR analyses by the same Δt (interval duration for bins) used in those calculations. The resulting values are not rates in the same sense as Foote's (2000) metrics. The rates $\hat{p}$ and $\hat{q}$ are derived based on a continuous process whereas CMR models births and deaths as discrete events, and the two variables do not scale with interval in the same manner. Nonetheless, rescaling the CMR probabilities by interval duration is useful as a means to directly compare the relative magnitudes of extinction and origination determined by these different approaches.

In the present context, it is important to note that the values of p, φ and γ are jointly and simultaneously estimated, using a likelihood function that incorporates all three parameters for each bin; variations in sighting (p) affect estimates of φ and γ for the same interval. Accordingly, CMR takes into consideration evidence for changes in sighting as part of the model fitting process. We employed the program FITMAN, written by one of us (HDS), to carry out the CMR analyses. A full accounting of the behavior and application of this method is given in Chen Xu et al. (2005b) and in Liow and Nichols (2010). Like the program MARK (Cooch & White 2019), FITMAN generates a series of models that are ranked based on Akaike's Information Criterion (Akaike 1973, Burnham & Anderson 1998), which balances data fit and model simplicity. Model ranking is expressed as the "relative AIC weight," and sums to 1.00 over the set of models compared. We utilized a set of models that range from maximally simple (p, φ, and γ are fixed over all intervals) to the maximally complex (all three parameters vary throughout the time series), as well as various intermediate combinations of parameters that were fixed or fully variable over time. Thus, we test explicitly whether the data indicate that sighting probability varied significantly through the interval of the LOME and affected estimates of E and O. Finally, because the species diversity in our dataset is relatively small, we employ the more appropriate, sample-size-adjusted version of AIC (AIC_c) (see Burnham & Anderson, 1998).

An additional feature of FITMAN is that it allows us to conduct a goodness-of-fit (GOF) test to ascertain whether the most complex model under consideration adequately describes the observed data. The observed deviance in the fitted model is compared to the distribution of the deviance values obtained from a series of Monte Carlo simulations in which the fitted model is used as a generating model. While this process is not perfect, it does allow detection of situations in which the model is missing substantial features of the original dataset, a method described by Pradel (1996). Among the critical model expectations is the requirement that all taxa have similar dynamical properties, which is essential since CMR employs the entire dataset to obtain singular estimates of p_i, φ_i, and γ_i for each interval. In the present case, we expected a priori (as described in the Introduction section) that the dynamical histories of the Diplograptina and Neograptina were distinctly different and therefore we conducted CMR modeling for each clade separately as well as in combination. We examined these three CMR models (all species combined, and the two species sets separated by clade) for the B1 and B2 binning schemes. Because CMR models require equal duration sampling intervals, the B3 set was inappropriate for this method and so we did not conduct CMR analyses of that dataset.

FITMAN employs nonparametric bootstrapping to numerically determine 95 percent confidence intervals (CI) around the parameter estimates for each bin. These CIs were estimated from parameter distributions obtained by resampling the original distribution of extinctions and originations within intervals to form 100 bootstrap sets.

2.4 Species Prevalence

Lastly, in addition to the per-bin estimates of sighting probability (p_i) obtained from FITMAN, we made use of the fact that the binned horizons in our HA composite retain a record of all the occurrences of each individual species through the study interval. This feature allows us to assess the prevalence of each species in each bin, bin by bin. By *species prevalence* (*sp*) we mean the proportion of horizons in a bin that contain a record of the species. We calculate a value of this metric for each species during every one of the bins within the species' inferred temporal range. Thus, for each species there is a time series of *sp* values that correspond to the interval in which it was extant. Similarly, for each bin there is a set of *sp* values representing the set of species that were extant during that time interval. As with the boundary-crosser metrics, we omitted from this analysis species that are confined to single bins.

Previous work has demonstrated that graptolite communities experienced substantial changes in community structure and species' relative abundances in the interval leading into the LOME (Sheets *et al.* 2016) and suggest that such changes may have presaged the mass extinction itself. Species prevalence offers a related measure, in this case gathered from the present global dataset, which we compared to species turnover and sighting rates to gain additional insights into the driving forces behind the LOME.

3 Results

3.1 Face-Value Turnover Metrics

Trajectories of species richness (estimated mean standing diversity, EMSD) obtained by the face-value boundary crosser metrics from three binning schemes are each very similar to one another (Fig. 5A) and differ greatly between the Diplograptina and Neograptina as well as among their subclades (Fig. 5B). Consistent with the finding of Goldman *et al.* (2011), the Neograptina are entirely absent from our samples prior to their appearance in the paleotropics during the Mirus Subchron near the end of the Katian Age. The peak in graptolite diversity that occurred during the Typicus Subchron within the Pacificus Chron consisted entirely of diplograptines (EMSD = 42.5 – 44.5, depending on the binning treatment); predominantly species of the Diplograptiodea and the Dicranograptiodea (see also Sadler *et al.* 2011, Cooper *et al.* 2014, Crampton *et al.* 2016). Climacograptoids had a relatively low and steady diversity (7.5–9 species) until late in the Pacificus Chron when they and the other two subclades began their decline toward final extinction. Crampton *et al.* (2016) noted that the LOME preferentially affected long-lived species to a greater degree than was the case during other times in the Ordovician, and this feature is reflected in our analysis as well: almost no diplograptines went extinct during the > 2 myr-long Complexus to early Pacificus interval that preceded the onset of the LOME (Fig. 5B). Similarly, except for a brief surge in the number of dicranograptoid species originations during the early Pacificus and Typicus subchrons, the number of diplograptine originations (Fig. 5 C) peaked early in the same pre-LOME interval in which there were nearly no observed extinctions. Diplograptine originations declined to zero by the start of the HICE in the latest Katian, just as the number of extinctions began to rise and the neograptines appeared in the paleotropics.

The appearance of neograptines during the Mirus Subchron slightly precedes the onset of the HICE in our scaled ordination and manifests as a pulse of species appearances among both the stem-neograptines and the Retiolitoidea within the latest Katian and early Hirnantian. Previous biogeographic analyses suggest that the appearance of Neograptina in the paleotropics at this moment represents an immigration event from mid- to high-latitude regions (Goldman *et al.* 2011). This immigration event coincides in our analyses with the largest number of diplograptine species extinctions recorded during the Late Ordovician (Fig. 5B,C). It is this dramatic disappearance of diplograptine species (and the consequent change in the composition of graptolite and other species assemblages) around the start of the Hirnantian Age that is commonly identified as the first pulse of the LOME. We will return to a discussion of the nature of that event later.

The mean standing diversity values for the graptolite clade as a whole fell to 17.5–15.5 species near the end of the Persculptus Chron; an approximately 62 percent loss of species diversity. The Neograptina diversified during the LOME and became the most specious clade concomitantly with the onset of the postglacial flooding event early in the Persculptus Chron, only some 600 kyr

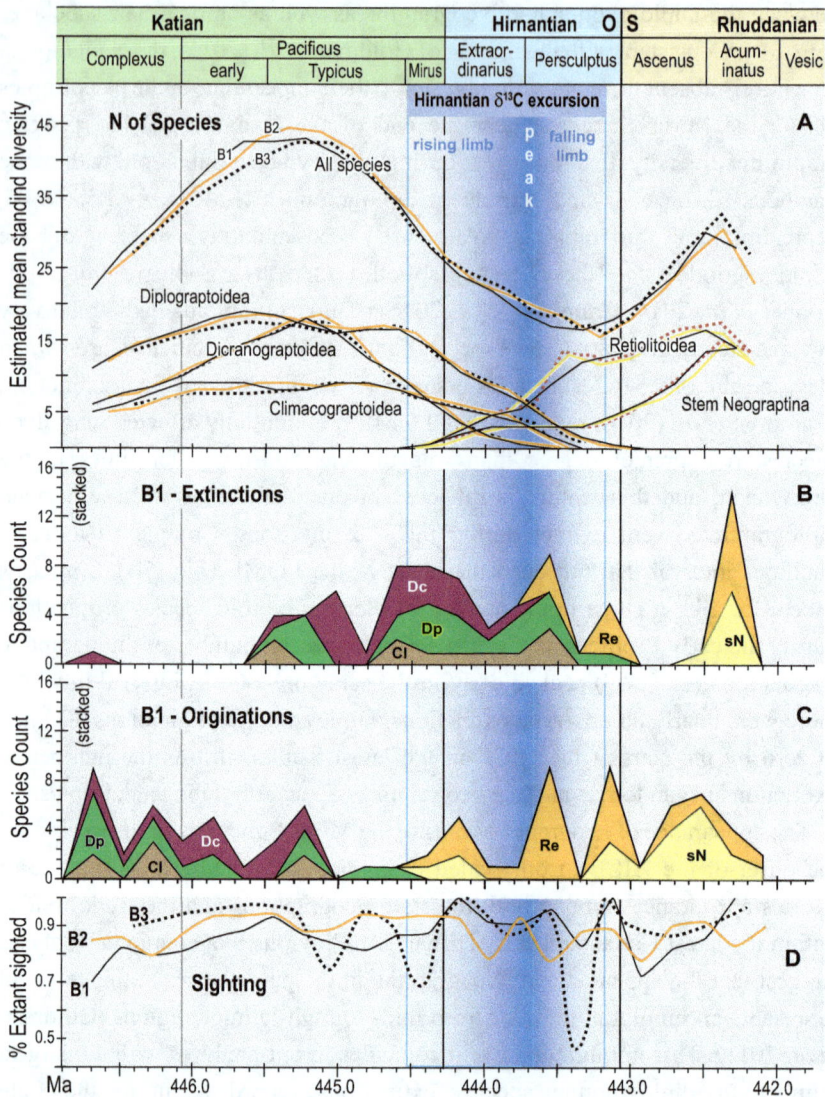

Figure 5 Observed diversity dynamics of planktic graptolite species through late Katian to early Rhuddanian chronozones and relative to the span of the HICE (as in Fig. 4). **A**, Estimated mean standing diversity in binning schemes B1–B3 for all graptolite species present through the study interval taken together, alongside those of the three constituent subclades within the Diplograptina: the Dicranograptoidea (Dc), Diplograptoidea (Dp) and Climacograptoidea (Cl), and two subgroups within the Neograptina: stem-group Neograptina (sN) and Retiolitoidea (Re). Note that variation among results from B1–B3 is small relative to the large changes in species diversity and

after their initial invasion of the paleotropics. By the early Silurian, in the Acuminatus Chron approximately 1.2 Myr from its nadir, graptolite EMSD had nearly recovered to Katian peak values.

All three of our binning schemes produce essentially the same pattern of change in estimated per capita extinction and origination rates (Fig. 6A,B). Among the Diplograptina rates of origination fall to near zero values by the end of the Typicus Chron while extinction rates slowly rise to a modest peak at the start of the Hirnantian, irrespective of bin boundary placement or whether the bins equalize sampling interval length or the number of horizons in each interval. This general similarity continues through the LOME, with the exception that the very short bin durations in the B3 set through the Katian–Hirnantian transition interval and the postglacial late Persculptus Chron interval produce exceptionally high $\hat{p}$ values among the Neograptina during both of those intervals (Fig. 6B) as well as an exceptionally high $\hat{q}$ among the Diplograptina in the late Persculptus Chron (Fig. 6A). The longer durations of the B1 and B2 bins recover similar peaks in the per capita rates but tend to spread them over a broader interval and in some cases shift their timing earlier or later by a half bin. In any case the two clades display strikingly different dynamical histories. This is most notable in the case of the late Persculptus peak rates: Diplograptines experienced high rates of extinction at precisely the same time that Neograptines speciated at high rates.

Caption for Figure 5 (cont.)

to the differences in those changes in diplograptine versus neograptine subclades. **B**, Stacked plot of the number of species extinctions within B1 bins by subclade. **C**, as in B but for species originations. High numbers of diplograptine species extinctions preceded the beginning of the HICE and the invasion of the paleotropics by neograptine species, which subsequently diversified while diplograptines went extinct over the course of the Hirnantian and earliest Rhuddanian. **D**, Time series of approximate per species sighting probabilities (proportion of observed, extant species recovered in bin) for each binning scheme; values are somewhat variable but are similar among binning schemes. Values show no long-term trend and those in the mass extinction interval (Mirus + Hirnantian bins) are not significantly different from nonextinction interval values; overall the sighting probabilities average 0.88 ± 0.22 (95 percent CI). Cl: Climacograptoidea; Dc: Dicranograptoidea; Di: Diplograptoidea; sN: stem neograptines; Re: Retiolitoidea.

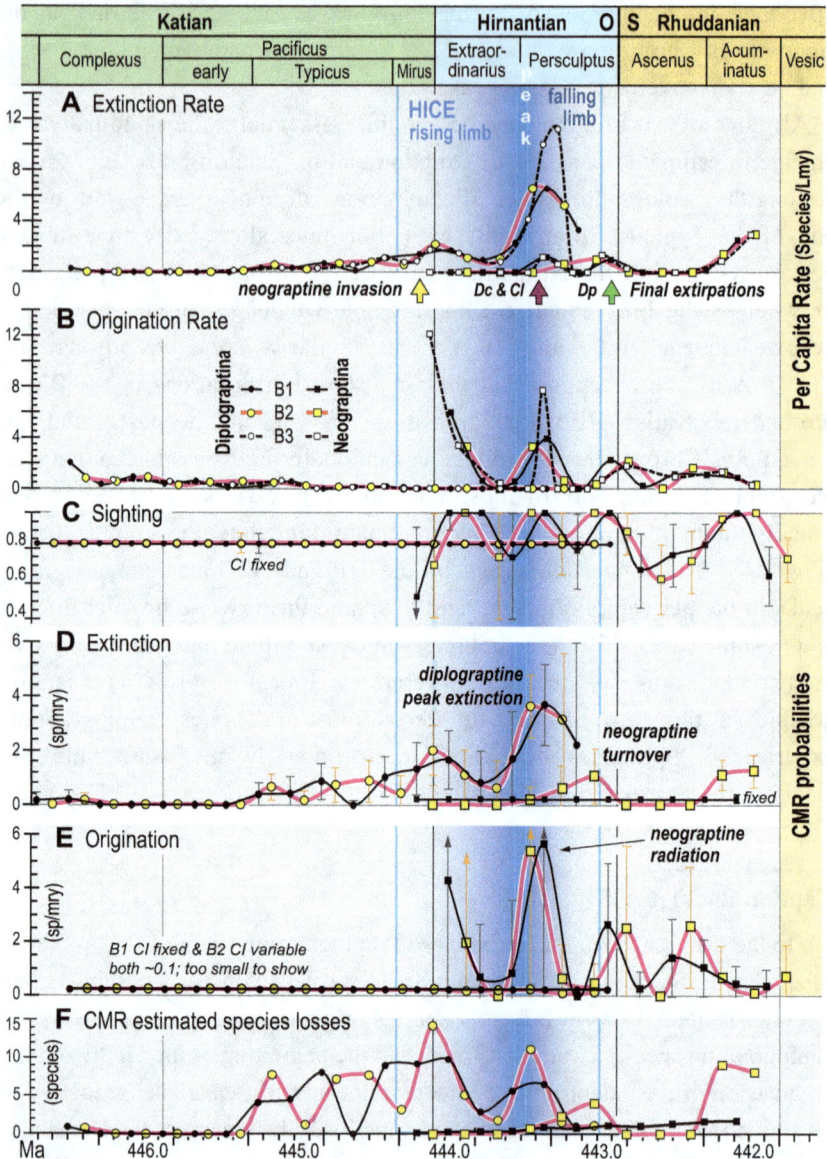

Figure 6 Time series of per capita and Capture-Mark-Recapture (CMR) model-based estimates of graptolite species turnover dynamics. Species of the clades Diplograptina and Neograptina analyzed separately based on the B1–B3 occurrence records for the per capita rates and the B1 and B2 records for CMR. Timing of neograptine invasion and diplograptine subclade final extinctions shown by arrows along the timeline below A (abbreviations as in Fig. 5). **A,** Per capita extinction rate ($\hat{q}$) from B1–B3 data treatments. **B,** Per capita origination rate ($\hat{p}$) from B1–B3 data treatments. **C,** CMR modeled species sighting

3.2 Capture-Mark-Recapture Models

We generated CMR models based on the B1 and B2 data treatments for 'all taxa' and for the Diplograptina and the Neograptina individually (Table 2). As we noted earlier, the B3 bin set with its unequal interval lengths is not suitable for CMR analysis. Each analysis included a set of models that range from maximally complex models in which all parameters are fully time-variant (p_t φ_t γ_t) to those that are fully time-invariant (p. φ. γ.) and ranked models according to their AICc weights. The all-taxa models fit parameters to the entire 24-bin time series, whereas those for the two clades employed a shorter time series corresponding to the duration of the clades: bins 1–18 for the Diplograptina and bins 13–24 for the Neograptina. In no case were fully time-invariant models the most highly ranked. For the B1-all taxa models, [p. φ_t γ_t] is the highest ranked model with sighting probability treated as fixed (Table 2). The corresponding B2 binning failed to pass the GOF test. The deviance value for the B2 all-taxa dataset was higher than that of 95 percent of the Monte Carlo simulations, indicating that those data are more poorly fit than expected by the most complex (fully time-variant model). As the per capita rates suggest (Fig. 6A,B), the Diplograptina and Neograptina display quite different turnover histories through the Late Ordovician and as a result the combined sighting record of the two clades in the B2 data treatment may not be sufficiently homogenous in its dynamical properties to be fit reliably by standard CMR methods. The best-fit models recovered for the two individual clades varied somewhat depending both on clade and binning. In the case of the Diplograptina, the highest ranked models for both the B1 and B2 data treatments are [p. φ_t γ.] with fixed rates of sighting and origination (Tables 2, 3). Only extinction rates varied significantly over time in these models of diplograptine turnover. In contrast, the most highly

---

Caption for Figure 6 (cont.)

probabilities (±95 percent bootstrapped CI); sighting rates and CI fixed for both of the highest ranked models of the Diplograptina record and variable for both Neograptina models. **D,** Extinction rates (±95 percent CI) derived from the highest ranked CMR models; rates time-variant for the Diplograptina in both models and only slightly variable or fixed for the Neograptina. **E,** Origination rates (±95 percent CI) derived from the highest ranked CMR models; rates fixed or minimally time-variant for the Diplograptina and highly variable for the Neograptina in both models. **F,** Number of species extinctions in the two clades inferred from the highest ranked B1 and B2 CMR models.

Table 2 Capture-mark-recapture model rankings. Results shown for five combinations of bin sets (B1 and B2) and clade-based data subsets: all species, Diplograptina only and Neograptina only. GOF p: p values for goodness-of-fit between the data and model expectations. Six alternative CMR models are shown for each of these datasets, with models ranging from fully time-variable (bottom row) through five combinations of fixed versions for model parameters p, ϕ, and γ (see text for parameter descriptions), with the subscript (t) indicating parameters that are variable among temporal bins and (.) those that are fixed over all bins. Model parameters shown are AIC_c, delta AIC_c (departure of AIC_c value for a particular model from the lowest AIC_c model in each set) and relative AIC_c weight (wt). The preferred models (italics) are those with lowest AIC_c score (and thus, zero delta AIC_c) and the highest relative AIC_c wt. See the text for further explanation.

B1 all-species			GOF p = 0.23	
	Model	AIC_c	Delta AIC_c	AIC_c wt.
	$p. \varphi_t \gamma.$	1607.59	13.10	0
	$p. \varphi_t \gamma_t$	*1594.49*	*0*	*1.00*
AIC Model choice	$p_t \varphi. \gamma.$	1633.71	39.22	0
	$p_t \varphi. \gamma_t$	1638.86	44.36	0
	$p_t \varphi_t \gamma.$	1614.21	19.72	0
	$p_t \varphi_t \gamma_t$	1605.95	11.46	0
B1 Diplo			**GOF p = 0.86**	
	Model	AIC_c	Delta AIC_c	AIC_c wt.
	$p. \varphi_t \gamma.$	*957.33*	*0*	*1.00*
	$p. \varphi_t \gamma_t$	963.51	6.19	0
AIC Model choice	$p_t \varphi. \gamma.$	1002.73	45.40	0
	$p_t \varphi. \gamma_t$	1006.12	48.79	0
	$p_t \varphi_t \gamma.$	967.44	10.11	0
	$p_t \varphi_t \gamma_t$	979.35	22.02	0
B1 Neo			**GOF p = 0.20**	
	Model	AIC_c	Delta AIC_c	AIC_c wt.
	$p. \varphi_t \gamma.$	495.82	22.24	0
	$p. \varphi_t \gamma_t$	484.80	11.22	0
AIC Model choice	$p_t \varphi. \gamma.$	475.62	2.04	0.02
	$p_t \varphi. \gamma_t$	*473.58*	*0*	*0.98*
	$p_t \varphi_t \gamma.$	482.37	8.79	0
	$p_t \varphi_t \gamma_t$	484.57	10.99	0
All-species model from above		1594.49	160.43	0
Combo Diplo & Neo B1 models		1434.06	0.0	1

Table 2 (cont.)

B2 Diplo			GOF p = 0.87	
	Model	AIC_c	Delta AIC_c	AIC_c wt.
	p. φ_t γ.	**923.02**	**0**	**0.99**
	p. φ_t γ_t	934.20	11.17	0
AIC Model choice	p_t φ. γ.	951.40	28.38	0
	p_t φ. γ_t	956.32	33.30	0
	p_t φ_t γ.	925.37	2.35	0
	p_t φ_t γ_t	938.50	15.48	0
B2 Neo			GOF p = 0.07	
	Model	AIC_c	Delta AIC_c	AIC_c wt.
	p. φ_t γ.	525.25	17.65	0
	p. φ_t γ_t	512.17	4.57	0
AIC Model choice	p_t φ. γ.	516.75	9.14	0
	p_t φ. γ_t	508.60	1.00	0.12
	p_t φ_t γ.	513.25	5.65	0
	p_t φ_t γ_t	**507.61**	**0**	**0.88**

ranked models of the turnover dynamics of the Neograptina were the fully time-variant model [p_t φ_t γ_t] for the B2 data treatment but, surprisingly, a very low and fixed risk of extinction [p_t φ. γ_t] for the B1 neograptine occurrence record (Table 2). In that model all the variation in species turnover is attributed to a combination of variation in the probabilities of origination and sighting.

Because the log likelihood values underlying the AIC are additive, it is possible to compare the fit of the B1-all taxa model to a combination of the separate Diplograptina and Neograptina B1 models (Table 2). The combined models yield a substantially lower summed AIC value (reported here following correction to AIC_c) than that of the highest ranked all-taxon model. Indeed, the resulting deltaAIC_c is larger than the difference between any of the alternate all-taxa B1 models (Table 2). This result indicates that the dynamical history of graptolites through the LOME is much better fit when the clades are modeled separately. Consequently, we focus for the remainder of the paper on the two-clade CMR models.

The estimated sighting probabilities of ~0.83 for diplograptines leads to higher projected total species diversity in the late Katian (52–58 species) than the estimated diversity obtained via the face-value metrics since in this formalism the observed species diversity is thought to underestimate true standing diversity by 1–p. Additionally, the CMR models suggest a slightly larger, 64–67 percent decline in species diversity during the LOME but recover a similar

Table 3 Cohort survivorship tables for diplograptine and neograptine species documenting highly significant extinction selectivity between clades in the two maximally different bin sets (B1 in A,B; B3 in C,D). **A, C** LOME-1 extinction selectivity; starting species cohort is the set of species present in the interval just before and during the LOME-1 extinction peak in the early Hirnantian and survivors are those still extant during some part of the interval up to and including the LOME-2 extinction peak in the mid Hirnantian. **B, D** LOME-1+2 extinction selectivity; starting species cohort is the set of species present in the interval from just before the early Hirnantian LOME-1 extinction peak up to and including the LOME-2 extinction peak, and survivors are those still extant during some part of the postpeak interval in the late Persculptus Chron and younger. Only one of the 31 diplograptine species in these cohorts survives both episodes in contrast to 16 of 18 Neograptine species.

A. (B1)	SURVIVE	EXTINCT	total cohort	% extinct
DIPLO	16	15	31	48.40%
NEO	7	0	7	0.00%
total cohort	23	15	38	
			two-tailed Fisher Exact p: 0.0291	

B. (B1)	SURVIVE	EXTINCT	total cohort	% extinct
DIPLO	2	29	31	93.50%
NEO	16	2	18	11.10%
total cohort	18	31	49	
			two-tailed Fisher Exact p: 6.026e-9	
			Chi-square p: < 0.0001	

C. (B3)	SURVIVE	EXTINCT	total cohort	% extinct
DIPLO	15	16	31	51.60%
NEO	8	0	8	0.00%
total cohort	23	16	39	
			two-tailed Fisher Exact p: 0.0125	

D. (B3)	SURVIVE	EXTINCT	total cohort	% extinct
DIPLO	1	30	31	96.80%
NEO	16	2	18	11.10%
total cohort	17	32	49	
			two-tailed Fisher Exact p: 7.325e-10	
			Chi Square p: < 0.0001	

early Persculptus Chron crossover in the diversity trajectories of the two clades, some 630 Kyr (three bins) prior to the final extirpation of the Diplograptina.

In general, all the per capita rates and CMR estimates of extinction intensity follow comparable trajectories; however, the two prominent peaks in extinction risk present in the face-value rates during the Katian–Hirnantian passage interval ("LOME-1 peak" hereafter) and during the early phase of the falling limb of the HICE (LOME-2 peak) are more subdued in the CMR results. This reflects the fact that the CMR models attribute some of the missing taxa to incomplete sampling rather than to extinction. Consideration of the 95 percent CI bounds on these probabilities reveals two additional salient features. Firstly, the LOME-1 peak was preceded by an interval of elevated extinction risk during the Typicus Subchron that is not significantly lower than that in LOME-1 peak. This extended period of accelerated extinction rates is matched in the CMR results by substantial numbers of implied extinctions through the latest Katian (Fig. 6 F). Secondly, peak values in the LOME-1 were not significantly different than those of the LOME-2 peak, although extinction risk in that interval was significantly elevated relative to that in the Typicus Subchron and relative to the low extinction rates that occurred near the end of the Extraordinarius Chron between these two Hirnantian pulses.

The clade-specific CMR models of species origination reveal a pattern that is less similar to that of the per capita, face-value interpretations in that they estimate a constant low level of origination among the Diplograptina, rather than the modest but declining $\hat{p}$ shown in the face-value rates, attributing fluctuation in the frequency of species appearances to incomplete sampling (Fig. 6E). Consequently, nearly all variance in occurrences is mapped onto variation in extinction risk (Fig. 6D).

In striking contrast to the diplograptine models, sighting and origination probabilities are strongly time-variant in the CMR models of neograptine turnover (Fig. 6 C,E) and extinction risk is either fixed or only slightly time-variant (Fig. 6D). The 95 percent CI bounds on neograptine extinction risk indicate that the two models differ significantly from one another only in the early Silurian (Fig. 6D), where we suspect that edge-effects (false range truncations) may have contributed to the inferred higher extinction in the B2 set. Origination, on the other hand, reveals two relatively high peaks with similar magnitude and age placement in both binning treatments (differing only in age by the ~ half bin offset). The first of these peaks in origination corresponds to the LOME-2 interval when the Diplograptina experienced the highest levels of extinction risk (Fig. 6D,E). The second represents a third LOME turnover pulse, more or less coincident with the beginning of the Silurian Period, during which the dominance of neograptine clades shifted from retiolitoids to stem-neograptines (Fig. 5B,C) and the final Diplograptina went extinct just prior to the end of the Ordovician (Fig. 6A). Note that during the Hirnantian the origination timeseries for the Neograptina is based on small numbers of species, which leads to high uncertainty in the modeled origination probabilities. The time series is also truncated at its start because we cannot estimate a reasonable value of origination associated with the paleotropical debut of the neograptines during the Mirus Subchron (indicated by a arrow in Fig. 6C) and it is likely that the high origination rate at that point in some measure reflects immigration rather than in situ evolutionary origination of the sighted species. Curiously, the final extirpation of the three diplograptine clades all occurred during intervals of relatively high origination probability among the Neograptina (arrows labeled "Dc & Cl," and "Dp" in Fig. 6A).

3.3 Species Prevalence

In our examination of the proportion of horizons occupied by extant species within each bin (species prevalence or *sp*), we utilized the B1 and B3 bin sets. In the former, all bins have a duration of ~210 Kyr and differ from one another in the number of horizons included. In contrast, each of the bins in B3 encompasses 24 horizons but have substantially different durations (Figs. 4, 7B). Despite these differences in approach, the median of the *sp* values calculated for the diplograptine species during each bin exhibits a significant decline in through the late Katian and Hirnantian for both datasets (B1: $r = 0.7118$, $p = 0.0003$; B3: $r = 0.7475$, $p = 0.0001$). In both sets, median *sp* reaches a local minimum at the end of the Mirus Subchron (Fig. 7A) and that minimum is followed by a strong rebound in the Extraordinarius Chron, where *sp* values are comparable to those early in the Pacificus Subchron.

The observed trend in median *sp* value through the late Katian appears to primarily reflect declining prevalence of individual species rather than a preferential loss of more prevalent species. Twenty-three diplograptine species have multiple occurrences in the late Katian and early Hirnantian in the B-1 bin set. For each of these species, we compared the average *sp* value among its occurrences in Katian time bins to that for Hirnantian samples and consider any difference less than 0.0 ± 0.04 (20 percent of the observed standard deviation in the data) to be no change in *sp* (Fig. 7C,D; this plot is based on B1 data but those from B3 produce equivalent results; given that agreement we did not examine the pattern in B2 data). Under that conservative criterion, prevalence values for only two Katian diplograptines increase in the Hirnantian and seven exhibit no change, whereas *sp* values for 13 species decreased from the Katian into the Hirnantian. The two-tailed binomial probability associated with this degree of biased outcome is 0.004 (narrower no-change cutoffs of ±0.02 and zero yield outcome ratios of 4:4:15 and 6:0:17, both of which also differ significantly from random expectations). Not surprisingly, species with higher average prevalence during the Katian were more likely to continue into the Hirnantian. We calculated the average species prevalence across all Katian bins for each species

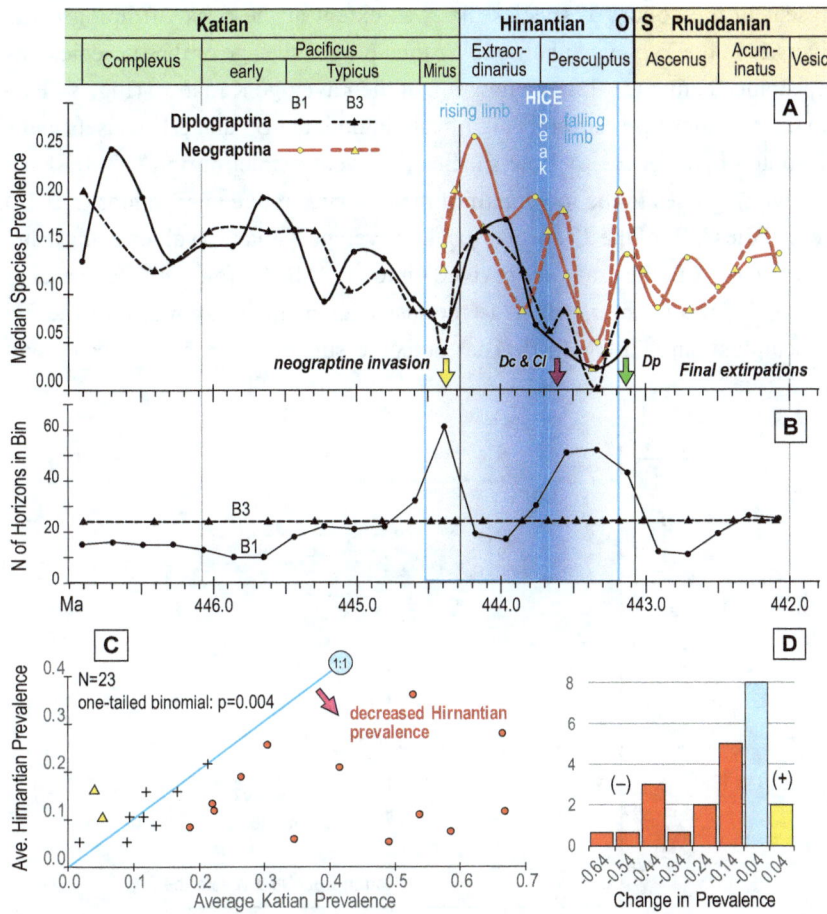

Figure 7 A, Time series of median species prevalence (the fraction of horizons within a bin that include a given species, assessed separately for each species) in the Diplograptina and the Neograptina derived from the contrasting B1 and B3 binning schemes. (B) Time series of the number of horizons per bin for B1 and B3, which have, respectively, intervals with a fixed 210 Kyr duration (B1, equally spaced midpoints) but variable sample sizes versus bins of variable duration (B3, unequally spaced midpoints) but with fixed sample size (24 horizons per interval). (C) Comparison of average species prevalence in diplograptine species during the Katian (ordinate) versus the average species prevalence of the same species during the Hirnantian; *Explanation of symbols*: (Δ), species with increased prevalence in the Hirnantian; (+), little change in prevalence (less than ±20 percent of Katian average); (•), decreased prevalence. (D) histogram of change in average species prevalence of diplograptine species based on values plotted in C. Values included in red bars (left of the mode) correspond to data plotted in C as (•), modal blue bar as (+) and right-most yellow bar as (Δ).

present in the cohort of species that was extant at the peak of diplograptine diversity in the Typicus Subchron. We then plotted the age of these species' last appearance datum (LAD) as a function of their average Katian *sp* (Fig. 8). Here too the relationship is strongly significant in both the B3 and B1 datasets and is illustrated here for B3. All the species that went extinct during LOME-1 had average *sp* values in the lower half of the observed range in *sp* values (< 0.35). Nevertheless, 19 of the 39 species that had average Katian *sp* values in that same lower half of the range survived into LOME-2 (or, in the case of *Paraorthograptus kimi*, which is the last known diplograptine, into the late Persculptus Chron LOME-3). That outcome suggests that even species with

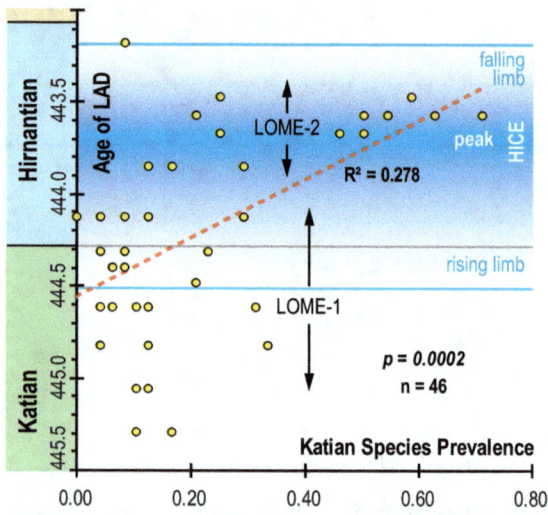

Figure 8 Correspondence (least squares regression, dashed red line) between the median prevalence of diplograptine species during the Katian and the age of their last appearance datum (bin midpoint) in the B3 dataset (B1 data yield the same result). The plotted species set is the cohort of 46 diplograptine species extant early in the Typicus Subchron (bin 5 of the B3 set), when the Diplograptina reached its peak diversity. The variance explained by the regression (R^2) is 0.278 and *p*, the probability of obtaining this relationship by chance, is 0.0002. Thirty-two species with low species prevalence in this cohort went extinct during LOME 1 (late Katian to earliest Hirnantian) but only seven such cohort members survived into LOME 2 (late Persculptus Chron) whereas all seven species in the cohort that had a species prevalence value in the upper half of the range (>0.32) survived into the later phases of LOME2. Accordingly, it appears that species' prevalence during the Katian significantly predicts their probability of survival during the LOME.

relatively low species prevalence during the Katian had a 50:50 chance of surviving LOME-1 and remaining among our sampled diplograptine taxa in the Hirnantian.

The preceding results could reflect sampling effects. In addition to the general higher rates of loss expected for rare taxa and backwards displacement of their apparent time of extinction (the Signore–Lipps effect, see Wang & Zhong 2018), sampling effects in this case could also reflect the loss of record caused by facies displacement or omission (e.g., Holland 2020, Zimmt et al. 2021). However, the local nadir in *sp* values during the Mirus Subchron coincides with an interval of relatively high sampling (large numbers of horizons in each B1 bin and short B3 bins) whereas the peak in median *sp* values during the Extraordinarius Chron occurs during an interval of lower sampling intensity (Fig. 7A,B). Because most species are uncommon (graptolite species included, Fig. 3; but see also Crampton et al. 2020), as one increases the number of horizons in a sample, the number of sightings for most species increases much more slowly than the number of horizons, which causes the median *sp* values in the B1 set to exhibit a weak but significant negative correlation with the number of horizons per bin (the slope of the least squares regression is ~0.05 but the 95 percent CI nonetheless excludes zero and the $R^2 = 0.36$ with a two-tailed p = 0.003). The residuals from this regression nonetheless exhibit a minimum in median *sp* during the same interval as for the original data. Furthermore, the B3 dataset has a constant number of horizons per bin and median *sp* values in that dataset are not significantly correlated with bin duration or other measures of sampling intensity in our data. Thus, the correspondence in the temporal trajectory of the median *sp* values in the B3 and B1 indicate that the decline in median *sp* through the latest Katian reflects a robust pattern of changing species prevalence. Additionally, this result is consistent with the community composition changes assessed from bulk samples at Vinini Creek (Sheets et al. 2016), where declining species diversity was accompanied by increased dominance of communities by a smaller number of species. As species diversity declined in this interval, many species were also becoming less common in graptolite collections, and presumably in graptolite communities.

The minimum in median *sp* during Mirus Chron coincides with the reinvasion of the paleotropics by the neograptines *Normalograptus angustus, N. ajjeri,* and *Metabolograptus ojsuensis*. Median *sp* values for the Neograptina at the time of their debut are higher than those of the contemporaneous Diplograptina. At Vinini Creek neograptine specimens, although representing relatively few species, are by far the most numerous elements of the early Hirnantian faunas (Sheets et al. 2016). The relatively high prevalence of neograptines is retained throughout the time series (Fig. 7A) and the two curves largely track one another in both the B1 and B3 sets. Following the recovery in median *sp* values during the early Hirnantian, values again decline precipitously at the time of the final extirpation of the three diplograptine clades during LOME-2 in the mid to late Persculptus Chron. The late Hirnantian minimum in species prevalence again occurs simultaneously in the B1 and B3 sets and again took place during an interval of relatively high sampling and low origination rates among the neograptines. These relationships suggest that low species prevalence may be causally linked to high extinction risk and reduced probability of origination.

One indication of the nature of that linkage comes from comparison of *sp* values for individual species with their patterns of specimen abundance in individual bulk samples. *Sp* values for individual species are significantly correlated with counts of specimen abundance. Sheets *et al.* (2016) documented shifting patterns of species abundance and community structure at two sites along the Laurentian margin of the Panthalassan Ocean (Vinini Creek and Blackstone River; Fig. 2). Bin-by-bin values of individual species prevalence are significantly correlated with the coeval specimen counts for those same species at Vinini Creek (Fig. 9A) and Blackstone River (Fig. 9B). The relations are illustrated here for B3 but the relationship in data from B1 is similarly significant, despite the considerable scatter in values present. Locally rare species (based on specimen counts) were generally uncommon globally as well, whereas species with locally larger populations tended to have correspondingly higher global prevalence.

The long-term decline in diplograptine species prevalence and the correlation of species prevalence with specimen abundance, together with the relatively higher species prevalence of contemporaneous neograptine species, accounts for the apparent abruptness of species losses through the Katian–Hirnantian transition at individual sections (Mitchell *et al.* 2007b, Sheets *et al.* 2016). Although precise specimen counts are available only for the Blackstone River and Vinini Creek sections (Sheets *et al.* 2016), qualitative abundance ranks are available for several sections in China (Wangjiawan, the Hirnantian global stratotype section, Honghuayuan, and Fenxiang; Chen Xu *et al.* 2000, 2005), Dob's Linn (the Silurian global stratotype section; Williams 1982) and Mirny Creek (Koren *et al.* 1983). In all these sections, the Diplograptina make up a substantial portion of number of recovered taxa in early Hirnantian assemblages (from a third to as much as 80 percent of the combined diplograptine and neograptine species diversity) but generally are much smaller components when the number of specimens of diplograptines is compared to that of neograptines (Fig. 10). At the intensively collected Mirny Creek section, *Parorthograptus* sp. occurs as one of only four taxa in the basal Hirnantian

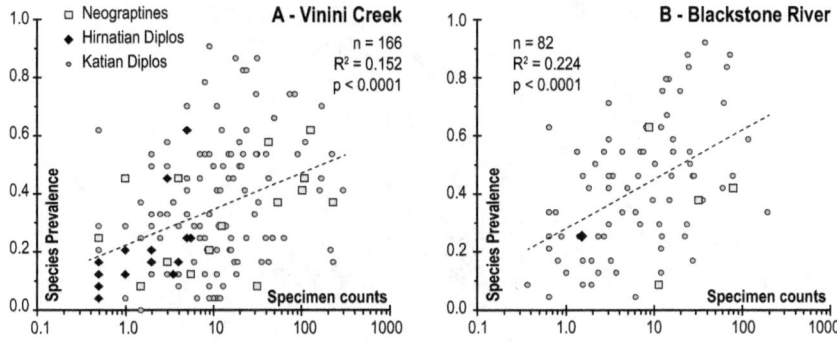

Figure 9 Regressions (dashed lines) between number of specimens of individual species in bulk samples (note log scale) and global species prevalence for those same species in coeval temporal intervals. Points labeled by taxon and age but regressions employed all taxa (n is the number of data pairs), with corresponding variance explained (R^2) and significance value (p). (A) species recovered in 16 bulk samples through the Vinini Creek section (mid Complexus to early Persculptus chrons) at Vinini Creek (Sheets *et al.* 2016) versus their coeval species prevalence values in temporal bins B3–2 to B3–16. (B), as for A but for bulk samples from Blackstone River (early Pacificus to early Extraordinarius chrons). The prevalence of species is significantly correlated with contemporaneous specimen abundance at each site, suggesting that global prevalence is a function, in part, of specimen density in local populations. Also note that in samples from the Hirnantian strata at Vinini Creek, the carry-over diplograptines generally have lower specimen counts and species prevalence than the diplograptine species did at that site during the Katian and are generally lower in both specimen abundance and species prevalence than the contemporaneous neograptines from that site.

strata there but specimens of this graptolite were so few and so poor that T. N. Koren' was initially reluctant to report it (personal communication, MJM; compare range charts in Koren *et al.* 1983, Koren' & Sobolevskaya 2008). At each of these sites, diplograptine specimens fall from 100 percent in assemblages during the mid to late Pacificus Chron to generally less than 20 percent (and often much less, especially in more near-shore sites) at the beginning of the Hirnantian and generally continue to decline in abundance through the Hirnantian more rapidly than diplograptine species diversity. Thus, although many diplograptine species survived the transition into the Hirnantian, they appear to have rapidly become minor components of Hirnantian graptolite communities throughout the paleotropics and are correspondingly difficult to recover in graptolite collections. Once again we see here the sort of common

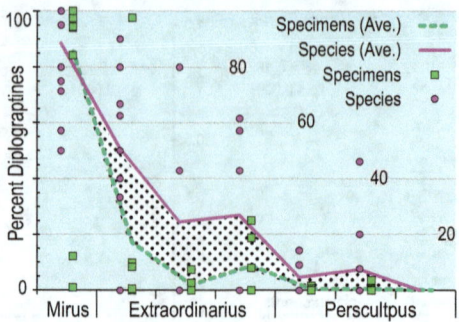

Figure 10 Time series of the percentage of faunas within individual Late Ordovician graptolite collections that are comprised by diplograptine species and diplograptine specimens versus the biozonal assignment of those collections. Data from Vinini Creek (site 24), Blackstone River (4), Wangjiawan North (25), Fenxiang (10), Honghuayuan (12), Mirny Creek (19) and Dob's Linn (6); binned by part of biozone: *D. mirus* Subzone, and lower, middle, and upper parts of the *M. extraordinarius* and *M. persculptus* biozones (see text for references). Late Katian assemblages were dominantly or entirely composed of Diplograptina but as a proportion of graptolite assemblages, the percent of recovered specimens that were diplograptine fell even more precipitously than did the proportion of diplograptine species in those assemblages (gap between average values indicated by stippled area).

cause linkage among specimen abundance, geographic range, stratigraphic duration and ease of collection – between ecology and geology. Notwithstanding that difficulty, it appears that the changes in *sp* documented earlier likely include a strong signal from changing graptolite metapopulation structures.

The alteration of graptolite community structure through the LOME is reflected in changing frequency distributions of species prevalence (Fig. 11). The illustrated distributions are derived from the B3 bin set, which once again yields results that are similar to the B1 set but are utilized here because they provide higher resolution through the Katian–Hirnantian transition than the equal duration bin sets. At the peak in graptolite species diversity during the Pacificus Chron (Typicus Subchron) histograms of species prevalence values exhibit the classic strongly right-skewed distribution of species abundance patterns: most species are relatively rare (have low *sp* values in this context, but as we demonstrated earlier, these values are significantly correlated with local species abundance, site by site), and the frequency of the more abundant species falls approximately exponentially (Fig. 11, Typicus-4–6). However, as

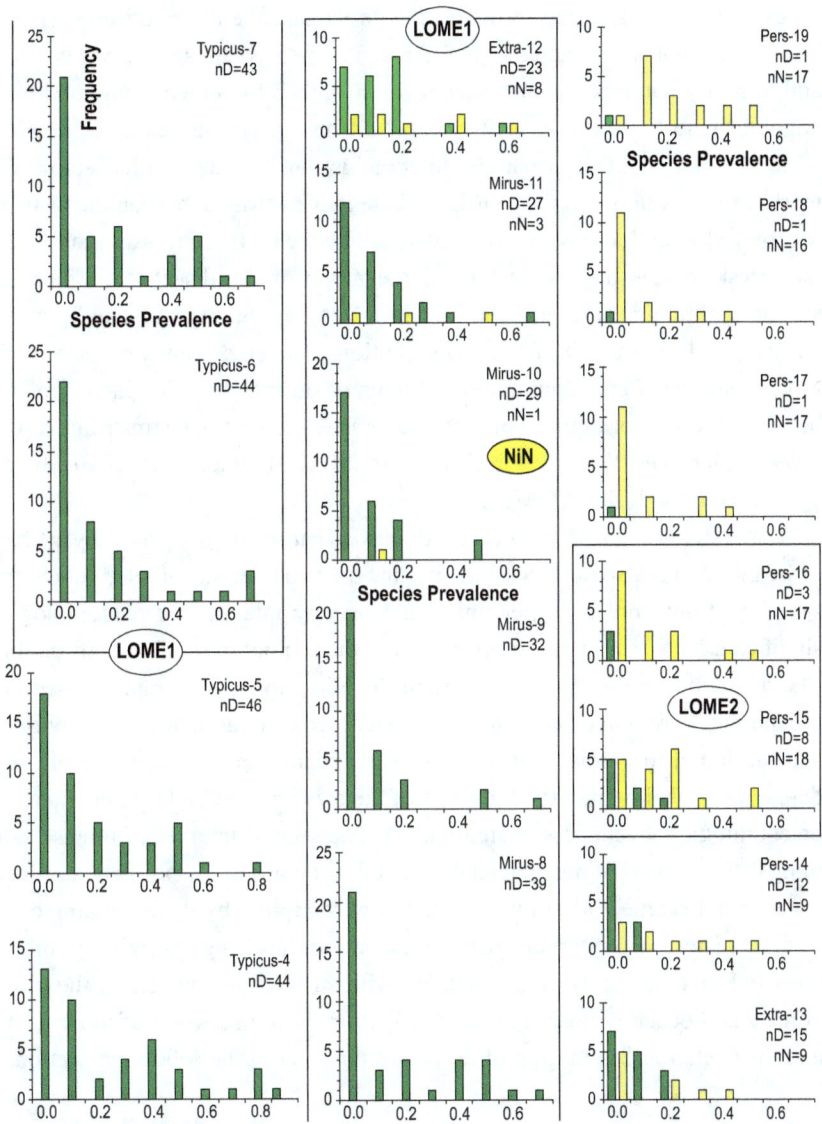

Figure 11 Interval-by-Interval (or bin-by bin) frequency distributions of individual species prevalence values for species present in each analytical interval (bin) beginning during the *Typicus* Subchron of the *Pacificus* Chron, when graptolite species diversity reached its Katian peak (bin B3–4) through the latest Hirnantian *Persculptus* Chron (B3–19); intervals labeled by chron and bin number. Prevalence values for diplograptines indicated by dark (green) columns and those of neograptines by pale (yellow) columns. Scope of the LOME-1 is indicated by the flattening and leftward shift (toward lower species prevalence,

species diversity declined through the late Katian, the distribution of species prevalence values contracted leftward and flattened as rare species went extinct and formerly common species became less so, thereby replenishing the low sp tail of the distribution (Fig. 11, Typicus-7–Mirus-9). This latter phase coincided with the interval of increasing extinction risk among the Diplograptina that marks the transition into the LOME-1. Late in the Mirus Subchron the distribution of diplograptine sp values flattened and by the early Extraordinarius Chron the rarest species no longer formed the mode of the distribution. The few surviving diplograptines and the invading Neograptina were unusually prevalent (Fig. 11, Extra-12). The sp distribution recovered somewhat across the Extraordinarius–Persculptus Chron boundary during the interglacial episode between the two principal Hirnantian ice sheet advances (Armstrong & Coe 1997, Melchin *et al.* 2013, Ghienne *et al.* 2014, Mauviel *et al.* 2020), before again flattening during LOME-2.

Although the changes in sp values described earlier are not primarily artifacts of sampling, it is also not a coincidence that the two intervals of relatively higher sampling (more horizons per unit time) occupy intervals of paleobiological significance (the neograptine invasion and the diplograptine final extirpation). The first encompasses the beginning of the Hirnantian, and graptolite sections around the world have been intensively sampled through that interval with the specific intent to precisely locate species appearances as part of the effort to define the global boundary stratotype of the stage and to better understand the paleobiological events associated with it. The second interval of intense sampling is the interval of postglacial flooding when graptolite-bearing black shales once again became widespread and is followed rapidly by the beginning of the Silurian. These two intervals also correspond to the rising and falling limbs of the HICE and so surely also coincided with substantial changes in the global climate and oceanographic conditions. We turn to a discussion of those events and their relationship to graptolite species turnover in the following section.

Caption for Figure 11 (cont.)

i.e., toward greater rarity) of the frequency distributions, which in the B3 set is exhibited by data from B3–6, late Typicus Chron through B3–12, early in the Extraordinarius Chron. Similar changes occurred during LOME-2 mass extinction phases in the late Persculptus Chron B3–15 and B3–16 intervals. nD and nN indicate the total number of diplograptine and neograptine species in each interval, respectively; NiN: indicates the interval (B3–10 Mirus Chron) during which the Neograptina invaded the paleotropics.

4 Discussion

4.1 Timing and Pace of Turnover

The present results indicate that the Late Ordovician change in graptolite species richness and taxonomic composition took place over an extended interval that, at the scale of our analyses, included three distinct turnover phases: first, during the interval of the Katian–Hirnantian boundary, then during the mid Persculptus Chron, postglacial flooding episode, and finally around the beginning of the Silurian (Figs. 5,6, 12; the apparent rise in extinction rate in the Acuminatus Chron is an artifact of range truncation at the end of the sampled time series). Chen Xu *et al.* (2005b), in their CMR analysis of graptolite turnover record in South China, recovered a similarly extended LOME-1 event as well as a strong turnover pulse near the end of the Hirnantian (LOME-3 as recognized herein). However, the LOME-2 event associated with the end of the postglacial flooding interval was obscured in the relatively poorly fossiliferous facies transition from the Kuanyinchiao lime mudstones (which contain the HICE peak in South China) into the overlying Longmachi black shales and so was expressed mainly as a pulse of neograptine origination and exaggerated extinction in the early Hirnantian. In contrast, analyses of a high-resolution, global composite by Crampton *et al.* (2016, 2018)[2] reveal an overall increased rate of graptolite species turnover during the Hirnantian punctuated by distinct peaks in origination and extinction rates that, as in our results, affect both rates simultaneously. Rather than three Hirnantian turnover peaks, the Crampton et al. results exhibit four, divided in the middle by a sharp decline in rates. We suggest that this lull in species turnover likely corresponds to the pause in turnover captured herein between the extended late Katian to early Hirnantian LOME-1 episode (and corresponding to the first two peaks in the Crampton et al. results) and the pair of intense turnover events (LOME 2 & 3 in our results) that took place during the Persculptus Chron. That interval of reduced turnover appears to correspond to the mid-Hirnantian interglacial (i.e., the interval of sea level rise between LOGC 2 and 3 of Ghienne *et al.* 2014; Fig. 1 herein, see also Li Chao *et al.* 2021, Jin Si-ding *et al.* 2024). This interpretation is

---

[2] In many respects the analytical approach of Crampton *et al.* (2016, 2018) is similar to ours in that they constructed a composite from graptolite-bearing sections from around the globe; however, they employed Constrained Optimization (CONOP9) to order species' first and last appearance data without any temporal binning of those data. This approach provides a continuous record of changing diversity and species turnover at the average temporal spacing of ordinal events in the time-scaled composite (~ 37 kyr) but does not directly take into account uncertainty in event placement and hence in the temporal uncertainty associated with the species origination and extinction. Crampton et al. calculated $\hat{p}$ and $\hat{q}$ at each ordinal level and employed a 250 kyr moving window to produce smoothed origination and extinction rate curves. Thus, the smoothed rate estimates in Crampton *et al.* were derived in a very different manner than the rates presented here but employ a similar temporal window for their calculation as our 210 kyr binned results.

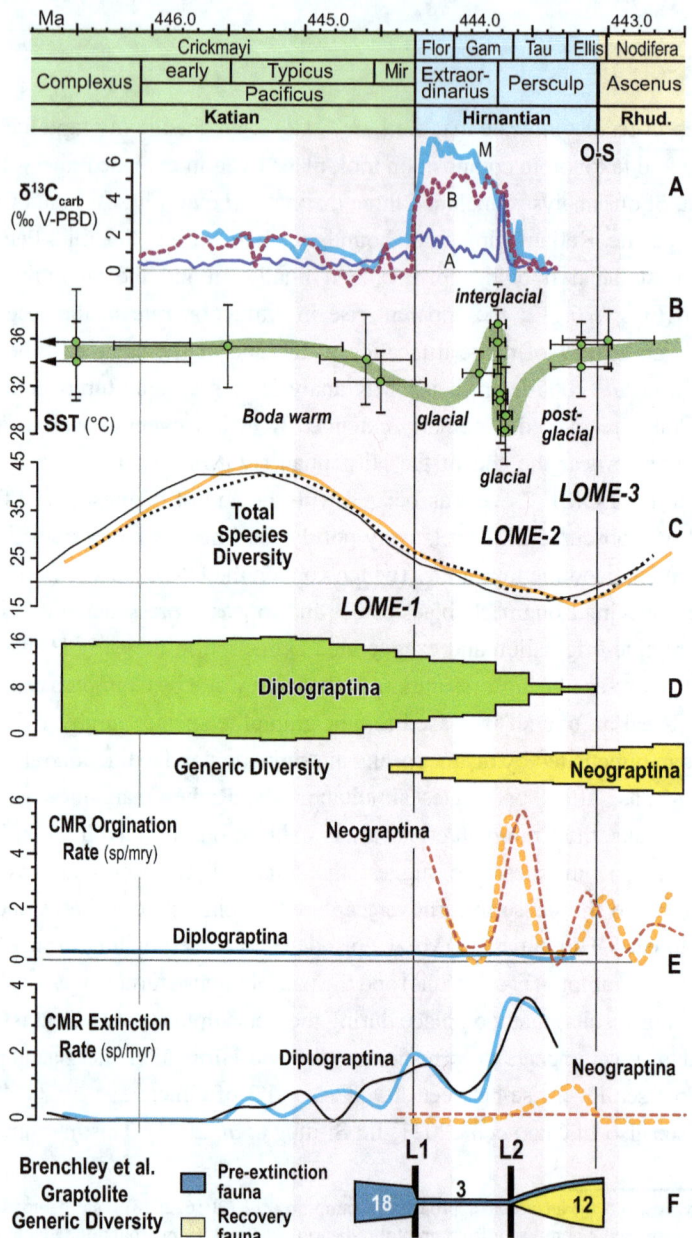

Figure 12 Comparison of time series of several measures of graptolite faunal turnover, including the three phases of the LOME described in the text (shaded horizontal bands), to those of estimated sea surface temperature, δ^{13}C, and to the Brenchley *et al.* (2001) model of graptolite generic turnover through the LOME. (A) δ^{13}C trajectories from Anticosti Island (narrow black curve "A"; from Mauviel & Desrochers (2016), Blackstone River (dashed curve "B"; from

supported by the spectral analysis of Crampton *et al.* (2018), which those authors concluded reflects orbital forcing of climate change and its effect on graptolite biotopes (see also Cooper & Sadler 2010, Cooper *et al.* 2014).

Brenchley *et al.* (2001) presented a quite different model of the LOME in which essentially the entire graptolite extinction took place in a geologically abrupt event (L1 in Fig. 12G) following which diversity remained low with no new taxa originating prior to the succeeding second pulse (L2) at the end of the Hirnantian glacial epoch. The L1 diversity decline was depicted as occupying a small fraction of the estimated 0.5–1 million years that separated the L1 and L2 events (Brenchley *et al.* 2001, p. 330 and their figure 2). Despite its striking disagreement with a prior quantitative study of graptolite turnover (Melchin & Mitchell 1991) and several subsequent publications (Chen Xu *et al.* 2005b, Fan Jun-xuan & Chen Xu 2007, Finney *et al.* 2007, Sadler *et al.* 2011), this interpretation has been employed explicitly in several recent studies (e.g., Hammarlund *et al.* 2012, Harper *et al.* 2014, Bond & Grasby 2017, 2020, Harper 2023) that relied on geochemical proxies to assess the potential causes of environmental changes during the LOME and implicitly in a large number of others (Bartlett *et al.* 2018, Zou Cai-neng *et al.* 2018b, Li Na *et al.* 2019, Hu Dong-ping *et al.* 2021, Li Na *et al.* 2021, Pohl *et al.* 2021, Jeon *et al.* 2022, Kozik *et al.* 2022b, Liu Mu *et al.* 2022, Qiu Zhen *et al.* 2022, Hu Rui-ning *et al.*

Caption for Figure 12 (cont.)

LaPorte *et al.*, (2009) and Monitor Range (thick blue grey line "M"; from LaPorte *et al.*, (2009) illustrating a range of different trajectories through the Hirnantian carbon isotopic excursion (HICE). (B) Sea surface temperature (SST) with 2σ uncertainty in SST (horizontal error bars) and estimated uncertainty in sample age (vertical error bars); SST data from Finnegan *et al.* (2011), but sample ages revised to reflect placement of the beginning of the Hirnantian Stage near the base of the Ellis Bay Formation at Anticosti Island (Achab *et al.,* 2011; Achab *et al.* 2013, Mauviel *et al.,* 2020; Zimmt & Jin, 2023; Zimmt *et al.,* 2024) and their correlation to samples in the Cincinnatian succession (Brett *et al.,* 2020; Sinnesael *et al.,* 2021). (C) Graptolite species diversity from Fig. 5. (D) Estimated mean standing diversity of graptolite genera (B1 bin set) in the Diplograptina and Neograptina. (E) Capture-Mark-Recapture estimates of species origination and extinction intensity for the Diplograptina and Neograptina, as in Fig. 6D,E. (F) The Brenchley *et al.* (2001) interpretation of graptolite generic diversity change through the LOME as presented in Harper (2023, Fig. 4), fit to the timing of the HICE.

2024, Liang Yu *et al.* 2024). Indeed, the LOME turnover event does appear as two abrupt extinction-only events, L1 and L2, as depicted in many of the stratigraphic sections examined in these studies. That appearance is largely an artifact, and the causes of the apparent abruptness are several. Holland and Patzkowsky (2015), Holland (2020) and Zimmt *et al.* (2021) have described the role of sequence stratigraphic effects on producing apparently pulsed extinction where none existed and we need not repeat that discussion here. We note simply that the ~200 m change in sea level and related facies displacements that were driven by the Hirnantian glaciation clearly fit the conditions under which sequence stratigraphic effects should be pronounced. The extinction pulses that have been identified at the majority of shelf sections, including several of those dominated by graptolite-bearing strata in Laurentia, South China, Chu-Ili Kazakh terrains, and Siberia (see Finney *et al.* 2007), coincide with sharp facies changes and these facies changes clearly have truncated many species ranges in these sections based on our current understanding of their global species ranges (see supplemental data files in Appendix B for range data). These facies and hiatus-related truncations contributed to the apparent abruptness of the LOME species losses in individual sections. Several particular features of graptolite ecology and biogeography also contribute to the apparent abruptness of the LOME turnover. First, diplograptines experienced a dramatic drop in global metapopulation sizes, which is reflected in widespread reduction in the abundance of diplograptine specimens in bedding plane assemblages (Mitchell *et al.* 2007, Sheets *et al.* 2016) and in their reduced species prevalence (Figs. 7,8,11). Most diplograptines seemed to go extinct in the latest Katian or early Hirnantian when they actually simply became rare (and harder to collect), even in otherwise abundantly graptolitic successions (Fig. 10). Those changes were accompanied by the immigration of several neograptine species that appeared widely in paleotropical assemblages around the globe (high species prevalence) and did so with relatively high specimen abundances (Goldman *et al.* 2011, Sheets *et al.* 2016), which turned most surviving diplograptines into relict species, effectively hidden among the abundant neograptines. This effect is especially strong at sparsely fossiliferous (e.g., Dob's Linn; Williams 1982, 1983)[3] or poorly productive sites (e.g., Mirny Creek; Koren *et al.* 1983), but is also evident even in highly productive successions, such as at Vinini Creek, where repeated intensive bulk sampling extended many species ranges compared to those

---

[3] This succession, represented in several faulted sections in the Dob's Linn valley contains only six graptolite-bearing black shale units through the entire *D. complexus* to mid *M. persculptus* zonal interval in the uppermost Hartfell Shale; merely one or two samples per biozone. Graptolite-bearing horizons become much more frequent in the overlying Birkhill Shale, which commences in the postglacial flooding interval of latest Hirnantian age and extends into the Llandovery.

initially reported there (compare Finney *et al.* 2007, Štorch *et al.* 2011) (see also the discussion in Mitchell *et al.* 2007b, Holland 2016, Sheets *et al.* 2016). Thus, the Brenchley *et al.* model mistakenly accepted the artificially abrupt turnover as an accurate reflection of the extinction dynamics. Recent work suggests that a similar revision of the trajectory of turnover among brachiopods may be necessary as well (e.g., Wang Guang-xu *et al.* 2019, Rong Jiayu *et al.* 2020, Zimmt & Jin 2023, Jin & Harper 2024). Whether the same will prove to be true of other clades during the LOME awaits further study.

Although the pace of change during the LOME clearly fluctuated dramatically, the picture that emerges from analyses of graptolite turnover indicates that the entire Late Ordovician glacial interval was a time of ecological and evolutionary flux. Declining species diversity and species prevalence together with rising extinction probability indicate that the LOME-1 events began in the late Katian (Typicus Subchron) as climate transitioned from the Boda warm toward the Hirnantian glaciation (Kröger *et al.*, 2019; Rasmussen *et al.*, 2019; Fan Jun-xuan *et al.*,2020; Deng Yi-ing *et al.*, 2021). This phase of the turnover reached a crescendo synchronously with the early Extraordinarius Chron glacial advance and carbon cycle disruptions recorded as the rising limb of the HICE and correlated changes in $\delta^{18}O$ due to increased continental ice volume reflected in lower global sea level. Overall the LOME-1 transformation of graptolite faunas encompassed an interval of nearly a million years duration, comparable to the scope of diversity decline and acceleration in extinction rates that affected Late Ordovician biotas more generally (Kröger *et al.*, 2019; Rasmussen *et al.*, 2019; Fan Jun-xuan *et al.*, 2020; Deng Yi-ing *et al.*, 2021). The duration of the LOME-2 and LOME-3 intervals, based on our data, again defined by the timing of final clade extinctions, species prevalence and extinction risk, was of the order of 200 kyr each. LOME-2 spanned the second major glacial advance from the Hirnantian interglacial into the succeeding postglacial sea level rise in the Persculptus Chron. The LOME-2 extinctions also accompanied the maximum extent of the carbon cycle disruption (recorded by the peak HICE excursion) and the interval of its rapid return to conditions more like those of previous epochs (falling limb of the HICE). As we discuss in more detail later, each of these extinction phases also involved new species originations among the Neograptina, again unlike the diversity trajectory depicted in the Brenchley et al. model, which only depicted extinctions (Fig.13C,D,G).

4.2 Selectivity of Turnover

The LOME turnover was highly selective at the clade level. Almost immediately upon discovery of the phylogenetic clade structure of planktic graptolites (Fortey & Cooper 1986, Mitchell 1987, Melchin 1998) it became apparent that the species turnover among graptolites involved a complete replacement of the formerly dominant Ordovician lineages by the greatly diversified descendants of a lineage that includes species of *Normalograptus* and their descendants (Mitchell, 1990; Melchin & Mitchell, 1991; Finney *et al.*, 1999; Finney, 2001; Chen Xu *et al.*, 2005b; Melchin *et al.*, 2011; Sadler *et al.* 2011). Survivorship of graptolite species cohorts through the LOME-1 and LOME-2 intervals demonstrate that extinctions throughout the LOME were significantly concentrated among the Diplograptina (Table 3). In contrast to the CMR models for the Neograptina, both of which include high origination rates (Figs. 5B,C), the best-fit CMR models for the Diplograptina posit a very low, albeit nonzero, origination rate (<0.3 sp/myr). Nevertheless, all known species originations during the Hirnantian occurred among the Neograptina, which (allowing for range uncertainty) added 16–19 species during the Hirnantian.

Unlike the LOME-1 and LOME-2 events, the LOME-3 episode affected only the Neograptina. By that point the Diplograptina were all but gone. Nonetheless, this event too, exhibited a substantial degree of selectivity. The initial radiation of neograptines in the early Hirnantian took place almost exclusively among the neodiplograptines (a paraphyletic stem group within the Retiolitoidea) and LOME-3 extinction occurred primarily among species of this clade whereas the stem-neograptines (which includes the conservative *Normalograptus* species) did not diversify until the postglacial Hirnantian (late Persculptus Chron) and early Silurian, when they gave rise to morphologically innovative dimorphograptids and monograptids (Fig. 5; Melchin *et al.* 2011, Bapst *et al.* 2012).

4.3 Simultaneous Origination and Extinction

The driving forces behind the LOME produced a positive correlation between origination and extinction throughout the mass extinction interval. Peaks in neograptine origination coincided with peaks in diplograptine extinction during both LOME-1 and LOME-2, each separated by an interval of low turnover near the end of the Extraordinarius Chron (Fig. 12E). Following their immigration from the cool-temperate zone into low-latitude regions at the outset of the LOME-1 (Goldman *et al.* 2011, Goldman *et al.* 2014), neograptine species richness (Fig. 5A), genus richness (Fig. 12D) and, to a lesser degree, structural diversity (Melchin *et al.* 2011, Bapst *et al.* 2012) expanded monotonically throughout the interval of the LOME. Thus, rather than being broadly detrimental for the pelagic biota (as one might expect for widespread oceanic anoxia, for instance), conditions during the LOME not only selectively favored the Neograptina, they did so contemporaneously (at least at the 210 kyr temporal scale of our dataset) with the negative effects that drove the Diplograptina to total extinction. In contrast, by the LOME-3 interval, diplograptines were all but extinct and the correlated peaks in origination and extinction in that episode almost exclusively arise from neograptine turnover.

4.4 Metapopulation Dynamics

The LOME turnover was accompanied by increasing rarity of diplograptine species. Diplograptines experienced significant reduction in species prevalence through the latest Katian and early Hirnantian (Figs. 7A,11) as well as a greatly diminished specimen abundance among the species that survived into the Hirnantian (Figs. 9,10). The result was that populations of most surviving diplograptines in the Hirnantian seldom reached local population sizes that were large enough to allow them to be routinely present in fossil collections. This suggests that diplograptine species occupied smaller geographic ranges and had more sparse populations, creating overall decreased total metapopulation sizes with lower spatial persistence. These changes in metapopulation dynamics exposed species to increased risk of extinction (Fig. 8). The immigrant neograptines replaced diplograptines during the earliest Hirnantian as the most widespread and abundant graptolite species. Furthermore, neograptines in general retained greater species prevalence and specimen abundance throughout the Hirnantian, regardless of the rise and fall of species prevalence in the two clades during the LOME (Figs. 7,9). This pattern of diminished diplograptine metapopulations is evident in well-studied paleotropical sites from around the globe; that is, from sections in South China (Wangjiawan-North, Hongjuayuan and Fenxiang; Chen Xu *et al.* 2000), western Laurentia (Vinini Creek and Blackstone River; Sheets *et al.* 2016) and Kolyma-Omolon (Mirny Creek; Koren *et al.* 1983), which together span ~ 210° in longitude around the paleotropics (Figs. 1, 2, 10).

4.5 Relationship to Changing Phytoplankton Communities

Drawing upon new and previously published N isotopic data, Melchin *et al.* (2013) argued that graptolite mass extinction during the LOME was strongly linked to changes in Late Ordovician phytoplankton communities. Evidence in support of this idea has continued to grow. The $\delta^{15}N$ of organic matter preserved in sediments deposited in the late Katian during the greenhouse conditions of the Boda interval immediately prior to the onset of the LOME generally have near zero values (±0.5 ‰). These low levels of ^{15}N-enrichment have been widely interpreted to be the result of nearly complete remineralization of fixed N by denitrification within the deep ocean in layers of persistently deoxygenated seawater, or in oxygen-depleted zones within areas of upwelling near the continental margins (LaPorte *et al.* 2009, Melchin *et al.* 2013, Luo Gen-ming *et al.* 2016, Liu Yu *et al.* 2020, Yang Xiang-rong *et al.* 2020, Yang Sheng-chao *et al.* 2021, Yang Xiang-rong *et al.* 2021). That interpretation suggests that photosynthesis in the relatively oligotrophic overlying water mass likely was dependent on biological fixation of N by cyanobacteria. This inference is supported by the fact that lipid biomarker suites from Boda interval samples are dominated by hopanes, consistent with those of cyanobacteria as well as 3β-methylhopanes (3-MeH) characteristic of methanotrophic proteobacteria (Rohrssen *et al.* 2013, Hu Rui-ning *et al.* 2024, Liang Yu *et al.* 2024). Considering the widespread occurrence of paleotropical diplograptines in sediments of that character, it is reasonable to suppose that these graptolites relied upon some combination of cyanobacteria and the chemosynthetic or heterotrophic microflora (including methanotrophic proteobacteria) of a tropical denitrification zone composed of oligotrophic shallow waters overlying expansive layers of anoxic deep waters (Finney *et al.* 2007, Cooper *et al.* 2012). In contrast, during the LOME interval, N isotopic data from low- latitude settings in Laurentia and South China suggest less intense denitrification concomitantly with an increase in sterane concentrations and other biomarkers consistent with green algal productivity as well as a reduction in the concentrations of 3-MeH, an indicator of methanotrophy (Rohrssen *et al.* 2013, Luo Gen-ming *et al.* 2018, Hu Rui-ning *et al.* 2024). These changes suggest that the deep waters of the oceans were more oxygenated during the glaciations, consistent with the reduction in methanotrophy and water column denitrification, and overall increased productivity among eukaryotic algal phytoplankton (Rohrssen *et al.* 2013, Shen Jia-heng *et al.* 2018, Hu Rui-ning *et al.* 2024) due to the increased availability of recycled sources of fixed-N (LaPorte *et al.*, 2009). The LOME-1 and LOME-2 intervals also included accelerated turnover in the fossil record of phytoplankton taxa in the paleotropics (Delabroye *et al.* 2011).

In this context it is worth recalling that eukaryotic algal cells are roughly an order of magnitude larger than bacterial cells, which are so small that molecular motion in sea water prevents individual cells from sinking. Shen *et al.* (2018)[4]

---

[4] The $\delta^{15}N$ data reported by Shen *et al.* (2018) from the Vinini Creek section include a significant (but previously unreported) correlation with lithofacies. $\delta^{15}N$ values from C-rich black shales in the Katian portion of the section are ~ 0.6 ‰ heavier than those from contemporaneous carbonate rocks. Consequently, the reported εTN_{por} values of the black shale samples are ~ 1.5‰ heavier than those from the carbonates. Because black shale samples dominated the Katian set but are entirely absent from the Hirnantian samples (reflecting the Hirnantian sea level fall and associated facies change at Vinini Creek and elsewhere), the apparent 1.5 ‰ shift in εTN_{por} at the base of the Hirnantian is an artefact of the facies change. Values based solely on carbonates exhibit an uninterrupted trend toward higher εTN_{por} and suggest increasingly more green algal-rich

argued that the increased productivity by algae during the late Katian and Hirnantian may have increased rates of carbon burial and, in turn, contributed to the C-cycle changes recorded in the HICE (we return to that topic later). Our point here is that this change to more abundant, larger phytoplankton is also likely to have negatively affected the food supply of paleotropical diplograptines.

Coeval strata from the cool-temperate Late Ordovician succession in the Prague synform (then part of the peri-Gondwanan microcontinent of Perunica) do not show a significant shift in $\delta^{15}N$ through the LOME interval, consistent with the relatively shallower, oxygenated and well-mixed water column in that region (Melchin et al. 2013). Under these conditions the Late Ordovician phytoplankton community in temperate and higher-latitude sites likely included a relatively high contribution from green algae compared to cyanobacteria and methanotrophs. Perhaps not coincidentally, all of the neograptine species known to have migrated into the paleotropics during the onset of the LOME-1 (*Normalograptus angustus, N. ajjeri, Neodiplograptus charis* and *Metabolograptus ojsuensis*) also occur in the Katian–Hirnantian succession in Perunica (Kraft et al. 2015, CEM unpublished), Iberia (Mitchell et al. 2011) and North Africa (Denis et al. 2007). Thus, the character of the phytoplankton community in low-latitude oceanic settings during the Hirnantian glaciation likely shifted away from the particular conditions to which the Diplograptina had adapted during the Boda interval and toward compositions more similar to those at higher latitudes; that is, toward conditions to which the Neograptina were already adapted.

These inferences are also consistent with those of Crampton et al. (2018). Their spectral analysis of a timescaled 64 Myr-long composite of graptolite species ranges (see also Footnote 2) revealed highly significant periodicities of 1.3 and 2.6 my duration in the probability of both extinction and origination. Based on comparison to the Earth's orbital periodicity during the Cenozoic, Crampton et al. concluded that these periods likely corresponded to Milankovitch "grand cycles" in obliquity and eccentricity, respectively. "These grand cycles would have modulated climate variability, alternating times of relative stability in the environment with times of maximum volatility, which influenced oceanic circulation and structure and thus, *phytoplankton populations at the base of the marine food web* (Crampton et al. 2018, p. 5686; emphasis added)," in agreement with the earlier studies

---

microfloras from the middle Pacificus Biozone into the Hirnantian without interruption at that site. This reinterpretation reinforces the conclusions presented in Shen *et al.*

of LaPorte *et al.* (2009) and Melchin *et al.* (2013). We return to the connection between graptolite turnover and climate cycles in the following section.

4.6 Relationship to C-Cycle and Anoxia

Considerable evidence indicates that anoxic to euxinic conditions were widespread during the latest Katian to Rhuddanian interval (e.g., Dahl *et al.* 2021, Li Na *et al.* 2021, Pohl *et al.* 2021, Kozik *et al.* 2022a, Song Li-cai *et al.* 2023, Yang Sheng-chao & Fan Jun-xuan 2025). Furthermore, several studies have argued from those data that the LOME extinction event resulted in whole or in part from the direct kill effects of anoxia itself or from sulfide or toxic metal poisoning, or some combination of these redox-related effects (e.g., Hammarlund *et al.* 2012, Vandenbroucke *et al.* 2015, Bartlett *et al.* 2018, Zou Cai-neng *et al.* 2018a, Bond & Grasby 2020, Dahl *et al.* 2021). Given that the case for anoxia as a driver of mass extinction hinges critically on the coincidence in time and space between anoxia and the extinction of the organisms it is alleged to have affected, we begin our consideration of the question of whether widespread anoxia may account for the timing and selectivity of the LOME turnover among graptolites with a brief synthesis of the temporal and spatial distribution of anoxia through the Late Ordovician and early Silurian.

The consensus of the available evidence suggests that ocean oxygenation through the latest Ordovician and early Silurian falls into three distinct phases. *1. late Katian (Pacificus Chron)*: anoxia was widespread, ranging from locally intermittent to persistent (Fig. 13; Melchin *et al.* 2013, Kozik *et al.* 2022b, Liu Mu *et al.* 2022, Song Li-cai *et al.* 2023). 2. *Hirnantian Glaciation (Extraordinarius – early Persculptus chrons):* in this interval anoxia was generally less persistent and more localized except during a brief period of sea level rise associated with the mid-Hirnantian interglacial warm period that separated the two main cycles of ice-sheet advance recorded in North Africa. The interglacial lies stratigraphically just below the main peak of the HICE around the globe, including several sections in South China (Li Chao *et al.* 2021, Li Na *et al.* 2021, Chen Qing *et al.* 2023), western Laurentia at the Vinini Creek, Blackstone River, and Monitor Range sections, (lower Persculptus Biozone, O5c max deepening; Fig. 1; Holmden *et al.* 2013, Sheets *et al.* 2016, Ahm *et al.* 2017) and eastern Laurentia on Anticosti Island (in the Lousy Cove Member of the Ellis Bay, with an occurrence of *Metabolograptus parvulus* and geochemical proxies suggesting global anoxia (Melchin 2008, Bartlett *et al.* 2018, Kozik *et al.* 2022a), and at Dob's Linn (the Extraordinarius Band, Fig. 1; Melchin *et al.* 2003, Bond & Grasby 2020, Mitchell & Melchin 2020), among other sites. 3. *Late Hirnantian to Rhuddanian*: anoxia became widespread again with the deglaciation of Gondwana and rapid sealevel rise, as indicated by profound shifts in geochemical redox proxies (e.g., Smolarek *et al.* 2017, Young *et al.* 2020, Dahl *et al.* 2021, Li Na *et al.* 2021) and by the global distribution of black shales with high TOC contents in deep shelf to basinal facies (e.g., Melchin *et al.* 2013, Yang Sheng-chao & Fan Jun-xuan 2025, and references therein). Melchin *et al.* recognized this particularly widespread interval of anoxia and high-TOC-bearing "hot" shales as the "Rhuddanian Oceanic Anoxia Event" (Rhuddanian OAE). The great preponderance of volcanic ash deposits (summarized in Longman *et al.* 2021) and Hg anomalies (Jones *et al.* 2017, Smolarek-Lach *et al.* 2019, Hu Dong-ping *et al.* 2021, Song Li-cai *et al.* 2023) that have been cited as evidence of voluminous volcanic activity in the late Katian to early Rhuddanian rocks occurs within the first and third of the three axoxia-rich intervals just outlined here. On the other hand, several studies have suggested that few of the reported Hg enrichments in this interval are volcanic in origin (Shen Jun *et al.* 2019, Shen Jun *et al.* 2022, Zhou Yu-ping *et al.* 2024) and thus it remains unclear to what degree volcanic activity may have contributed to the widespread development of anoxia during the Rhuddanian OAE.

Does the pattern of graptolite turnover during the LOME conform with the timing and expected effects of widespread, shallow oceanic anoxia? We believe it does not. It is conceivable that reduction in population size and fission of species ranges might well accompany increased oceanic anoxia or upwelling of sulfidic water masses and that these changes might increase the risk of extinction and the likelihood of species vicariance, and thus contribute to the linked extinction and origination observed among graptolites. The timing of widespread anoxia fits poorly to the timing of LOME-1 graptolite species losses, however. Firstly, diplograptines were at their peak diversity during the time of widespread Pacificus Chron anoxia; secondly, the rise in rates of extinction and falling origination among these species preceded and persisted longer than the pulse of anoxia locally present at the Katian–Hirnantian boundary; and thirdly, the highest rates of graptolite species extinction during LOME-1 occurred in the early Hirnantian when the oceans were less anoxic than previously. In contrast, the timing of anoxia during the late Hirnantian–early Rhuddanian postglacial interval (Rhuddanian OAE) does correspond well to those of LOME-2 and LOME-3 episodes, during which the few remaining diplograptines and numerous neograptines went extinct. We note that our results indicate that the duration of the species turnover during the Rhuddanian OAE was much more extended than the concentrated episode that is typically shown for LOME-2. Furthermore, the Rhuddanian OAE coincides with a marked diversification (both in taxa and in colony structure) among the Neograptina; an expansion that continued through the entire Rhuddanian Age (e.g., Melchin *et al.* 2011,

Sadler *et al.* 2011, Deng Yi-ing *et al.* 2021, Štorch 2023). Based on our understanding of the paleoecology of graptolites (see Cooper *et al.* 2012), however, we can envision no reasonable scenario in which these physical drivers could cause extinction to be limited almost exclusively to the Diplograptina and origination to occur only, and simultaneously, in the Neograptina, as evidently was the case, nor to produce the different turnover histories exhibited by the two main neograptine clades (stem-neograptines and the Neodiplograptidae).

It is important to keep in mind that diplograptine and neograptine species occur together in the great preponderance of latest Katian and early Hirnantian samples and so they must have occupied more or less the same sites and water masses most of that time. The same is also true of the two neograptine groups, species of which occur side by side in Hirnantian samples all across the paleotropics. Additionally, although the neograptines for which Sheets *et al.* (2016) inferred biotope affinities were all identified as epipelagic species, so too were the majority of the diplograptines and both they and the deeper-water, mesopelagic diplograptines ultimately all went extinct, in spite of the differences in their biotope affiliation and the initially higher rates of loss among the mesopelagic species. Thus, although a depth-stratified model of risk in which deeper-dwelling species were more affected by anoxia than the epipelagic species could account for the preferential species losses among mesopelagic species (see also Li Na *et al.* 2021), it fails to account for the overall joint pattern of graptolite origination and extinction.

An alternative explanation for the initially higher rate of extinction of mesopelagic species may lie in their resource utilization and likely smaller geographical ranges. In their study of the influence of biotope on extinction risk among Ordovician graptolites, Cooper and Sadler (2010) found that mesopelagic species had significantly and consistently higher rates of extinction. Since graptolites were most abundant and diverse around the margins of continental shelves (like microzooplankton today and oceanic productivity more generally), Cooper and Sadler reasoned that, other things being equal, species confined to deeper shelf-edge habitats likely had smaller geographic ranges (and so, overall smaller metapopulation sizes). They also noted that these more limited habitats likely were more sensitive to climate disruption. Boyle *et al.* (2017) also found that off-shore obligate graptolite species also had significantly higher extinction risk than did species that displayed no facies-limited occurrences but found that overall commonness (the combination of number of occurrences and geographic range), which again is likely to be related to total metapopulation size, provided the strongest explanation for variation in species extinction risk. Those inferences have received support from subsequent work that shows graptolite species turnover to be closely correlated with climate fluctuation and the related habitat disruptions (Cooper *et al.* 2014, Crampton *et al.* 2016, 2018). These features naturally would result in an earlier and more intense response among mesopelagic species, as a group, to the climate disruptions that accompanied the Hirnantian glaciation than the, on average, more widely distributed epipelagic species, consistent with the findings of Sheets *et al.* (2016) and Li Na *et al.* (2021).

Rather than direct kill effects, it appears that the principal importance of anoxia during the LOME interval is that it attests to the disruptions experienced by the oceanic ecosystem. Anoxia in this interval has been widely attributed to high rates of organic matter (OM) productivity and, together with the HICE, has been taken to record an interval of relatively high rates of OM burial relative to inorganic C sequestration rates (e.g., Marshall & Middleton 1990, Saltzman 2005, Qiu Zhen *et al.* 2022, Algeo & Shen Jun 2023, Young *et al.* 2023, Liang Yu *et al.* 2024). Several lines of evidence suggest that the surface waters of the oceans during the LOME-1 interval experienced an increase in nutrient supply (fixed nitrogen and P, among others) through invigorated upwelling (LaPorte *et al.*, 2009; Melchin *et al.*, 2013; Luo Gen-ming *et al.*, 2016; Liu Yu *et al.*, 2020; Yang Sheng-chao *et al.*, 2021; Zan Bo-wen *et al.*, 2024), higher volcanic weathering inputs (Yang Sheng-chao *et al.*, 2019; Longman *et al.*, 2021) and increased terrestrial runoff more generally (Lenton *et al.*, 2016; Porada *et al.*, 2016; Dahl & Arens, 2020; Müller *et al.* 2024). Ventilation of the deep ocean reduced the distribution of anoxic marine sediments and the associated burial of OC in sediments underlying anoxic marine bottom waters. To offset the reduction in anoxic marine facies, OC burial would need to have increased in oxygenated marine sediments to account for the HICE. Such an increase in OC burial could have been a result of the increased abundance of larger and more decay-resistant particles contributed from nonvascular land plants and eukaryotic algae (Lenton *et al.*, 2012; Parnell & Foster 2012; Rohrssen *et al.*, 2013; Lenton *et al.*, 2016; Shen Jia-heng *et al.*, 2018; Hu Rui-ning *et al.*, 2024), or by overall slower organic matter decomposition rates in the colder Hirnantian oceans (Stanley, 2010; Finnegan *et al.*, 2012a). Melchin *et al.*, (2013) noted, however, that there is no direct evidence of increased sedimentary carbon storage during the early mid Hirnantian: nearly all rocks of that age have lower concentrations of organic matter than those before and after the HICE rather than the predicted increases (see also Yang Sheng-chao & Fan Jun-xuan, 2025). Arguments that the missing carbon-rich sediments were largely confined to sediments deposited in the abyssal depths and subsequently subducted with the oceanic sea floor or destroyed by tectonic forces, although possible, seem forced at best considering that most oceanic productivity occurs near continental margins, and that bottom-water anoxia was more widespread before and after the main glacial epoch when sedimentary TOC contents were higher and $\delta^{13}C$ values were lower than in sediments deposited during the HICE. Furthermore, it is not clear how widespread anoxia in the abyssal realm could account for the LOME in any case. The observed diversity changes that constitute the LOME are virtually all based on the fossil record of organisms that lived on the continental shelves and in epicontinental seas rather than in the abyssal ocean floor, which has an extremely limited fossil record during the Paleozoic Era. Accordingly, it appears unlikely that the positive $\delta^{13}C$ excursion can be attributed to increased productivity or its burial efficiency (Olivarez Lyle & Lyle, 2006).

An alternative resolution to this apparent conflict may be found among the other factors that influence the $\delta^{13}C$ value of ocean C reservoir, such as the globally averaged $\delta^{13}C$ value of the weathered carbon flux from the continents, and the photosynthetic fractionation factor (ε_P) associated with marine primary production. Kump *et al.* (1999) and Kump and Arthur (1999) employed a simple box model to explore the implications of global changes in these parameters as an alternative means of forcing the HICE without increasing the fraction of organic carbon relative to inorganic carbon buried in marine sediments. Increasing the $\delta^{13}C$ value of weathered C flux to the oceans by 4‰ and ε_P by 7‰ ($\equiv \Delta^{13}C = \delta^{13}C_{carb} - \delta^{13}C_{org}$) could account for the magnitude of the largest reported C isotopic excursion (CIE) of 7‰, which they (and many others since, e.g., Brenchley *et al.*, 2003; Bergström *et al.*, 2009; Lenton *et al.*, 2016; Algeo & Shen Jun, 2023; Zhang Jun-peng *et al.*, 2024) assumed represented the global shift in the $\delta^{13}C$ values of the ocean and atmosphere during the glaciation. Subsequently, shelf gradients in the magnitude of the HICE were found by Melchin and Holmden (2006) and LaPorte *et al.* (2009): shallow nearshore environments generally record larger excursions than deeper offshore ones (see also Jones *et al.*, 2020). CIEs in offshore settings average just 2.7 ± 0.4‰ (LaPorte *et al.*, 2009), or 1.5‰ if only the deepest setting is considered (LaPorte *et al.*, 2009; Ahm *et al.*, 2017). These offshore sites are more likely than nearshore settings to record the true change in ocean C cycling during the glaciation. Additionally, many Yangtze Platform sections yield similarly low-amplitude HICE excursions (e.g., Gorjan *et al.*, 2012; Yan De-tain *et al.*, 2015; Zou Cai-neng *et al.*, 2018a; Song Li-cai *et al.*, 2023; Yang Sheng-chao *et al.*, 2024). Because the magnitude of the excursion in $\delta^{13}C$ is smaller than previously thought, the sea level–driven perturbation in carbonate weathering is by itself more than enough to account for the positive excursion. The 7‰ additional increase in globally averaged ε_P (Kump *et al.*, 1999), which subsequent studies failed to corroborate (Melchin & Holmden, 2006; LaPorte *et al.*, 2009); is not needed. Using the same model setup and C-cycle parameters as Kump et al. (1999), a 6 percent increase in carbonate weathering predicts a CIE of 1.5‰. If we instead use the full 22 percent increase in carbonate exposure inferred from geological maps, then the same 1.5‰ excursion can be achieved with a co-occurring decrease of 6 percent in the fraction of organic carbon buried relative to inorganic carbon buried in marine sediments (Table 4). In this scenario, the $\delta^{13}C$ value of the riverine flux of dissolved C to the oceans changes from –7 to –5‰ representing contributions from carbonate weathering (80 percent at 0‰) and kerogen weathering (20 percent at –25‰).

Considering factors that can change the $\delta^{13}C$ value of the continental weathering flux rather than the burial efficiency of organic carbon during the glaciation has the advantage of being able to explain the origin of the positive CIE without contradicting the evidence for low TOC contents of sediments deposited during the Hirnantian sea-level lowstand. It is also consistent with sequence stratigraphic and geochemical evidence that positive CIEs through the Late Ordovician and early Silurian commonly correspond to sea-level lowstands (Ludvigson et al., 2004; Saltzman & Young, 2005; Fanton & Holmden, 2007; Immenhauser et al.,

Table 4 Model parameters demonstrating 1.5‰ origin of the Hirnantian CIE by increased carbonate weathering during the glacio-eustatically controlled sea level lowstand. Initial C-cycle parameters are those of Kump et al. (1999). All parameter changes represent steady-state conditions. Subscripts sw, riv, w-carb, and org represent seawater, rivers, carbonate weathering, and organic matter, respectively. The following simplifying assumptions apply: $\delta^{13}C_{sw}$ is equivalent to sedimentary $\delta^{13}C_{carb}$; ε_P is the global average value of the photosynthetic fractionation factor, which is equivalent to $\Delta^{13}C$ (= $\delta^{13}C_{carb} - \delta^{13}C_{org}$); $\delta^{13}C_{riv}$ is the value of the continental weathering flux of carbon to the oceans, which is comprised of two sources: carbonate weathering (0‰) and kerogen weathering (–25‰), with f_{w-carb} being the fraction of carbonate-derived carbon in riverine C-flux. The modeled scenario is for the maximum change in f_{w-carb} based on geological maps depicting the area of exposed carbonates during the sea-level lowstand (Kump et al., 1999). A reduction in the fraction of organic carbon burial is required to keep the Hirnantian CIE from climbing above 1.5‰. Altered forcings are highlighted in underlining and responses in italics. See the text for discussion.

Interval	Oceanic C Reservoir		Weathering		Burial	
	$\delta^{13}C_{sw}$	ε_P	f_{w-carb}	$\delta^{13}C_{riv}$	f_{org}	$\delta^{13}C_{org}$
Pre-HICE	1.1	–31	0.72	–7.0	0.260	–31.0
HICE	*2.6*	–31	**0.80**	*–5.0*	**0.244**	*–28.4*
Post-HICE	1.1	–31	0.72	–7.0	0.260	–31.0

2008; Holmden *et al.*, 2013; Husinec & Leslie, 2022; Zimmt *et al.*, 2024). These considerations further suggest that increases in carbon burial and water column anoxia during the early mid Hirnantian may have been less extensive and persistent than previously supposed. Our argument here is that productivity during this interval shifted in character rather than increased – that productivity generally became both more localized and more variable in response to the climate variability that periodically altered local supplies of P, fixed N, and water column oxidation, and that these changes led to green-algal dominated phytoplankton communities that diplograptines were poorly equipped to utilize but that fit the needs of the exaptively well-equipped Neograptina.

The widespread anoxia associated with the rise of sea level concomitant with the retreat of the Hirnantian ice caps reflects the return to warm, stratified, oligotrophic oceanic conditions. Presumably those changes reshuffled the ecological deck yet again and resulted in the LOME-2 and LOME-3 turnover pulses (Finnegan *et al.*, 2012b; Melchin *et al.*, 2013; Harper *et al.*, 2014; Young *et al.*, 2020; Yang Sheng-chao *et al.*, 2021; Lu Yang-bo *et al.*, 2022; Liang Yu *et al.*, 2024; Zhang Jun-peng *et al.*, 2024). The linkage between anoxia, $\delta^{13}C_{DIC}$, nutrient-related productivity changes, and species turnover during the LOME is supported by the correlation between the positive excursions in C and N isotopes that occur during glacial intervals in general and during the LOME interval in particular (reviewed in Algeo & Shen, Jun 2023). In the latter instance, $\delta^{15}N$ values exhibit patterns of spatial variation among regions and among oceanic settings (relatively restricted shallow versus relatively open deep sites) that correspond with those of $\delta^{13}C$: large local changes in N cycling drove large local changes in C-cycling and isotopic fractionation that accentuated those of the global reservoir.

As we have argued earlier, the long-term record of graptolite turnover indicates that turnover among these zooplankton was intimately connected to fluctuations in the C-cycle (Cooper *et al.* 2014) in step with astronomically forced climate variability (Crampton *et al.* 2018). Interestingly, several sites that exhibit abrupt changes in $\delta^{13}C$ and $\delta^{15}N$ (e.g., Blackstone, Vinini Creek, see Figure 1) also exhibit abrupt changes in graptolite community composition, whereas at Wangjiawan, where the C and N stable isotope curves (Luo Genming *et al.* 2016) exhibit extended rising limb and a brief peak in the lower *M. persculptus* Biozone interval also shows an extended turnover in the graptolite faunas (Fig. 1). Nanbazi and Dob's Linn also exhibit a slow rising limb of the HICE through the *M. extraordinarius* Biozone strata with a distinct peak confined to the lower part of the *M. persculptus* Biozone there (Fig. 1) but unfortunately the Extraordinarius Chron interval at those sites produced few graptolite-bearing collections. Although it is possible that this pattern might be due in part to sedimentary condensation or hiatuses, we have found no independent sedimentological evidence for such at Blackstone or Vinini Creek, the latter of which also possesses a graptoliferous and apparently complete *M. extraordinarius* Biozone succession. Biomarker data, although also very sparse at this point, seem to follow a similar pattern – tracking rates of change in paleocommunity proxies and in N and C stable isotopes. Further studies are needed to test these relationships, but if these patterns are not artifacts, then it appears that local variation in the rate and timing of changes in N and C cycling may also have driven the pace of local population changes in plankton communities generally, including graptolites and the primary producers upon which they depended. We expect that these changes in N and C cycling likely affected many other organisms similarly.

4.7 Hirnantian Turnover Pulse Hypothesis

Building on her extensive work on Neogene mammalian faunas across Africa, Vrba (1985, 1993, see also Barnosky 2001, Faith & Behrensmeyer 2013), proposed the Turnover Pulse Hypothesis (TPH). She argued that alterations in habitat conditions brought about by Milankovitch cyclicity and climate change, especially during major episodes of cooling, led to pulses of migration, vicariance, and altered competition, which then led to turnover pulses – intervals of high origination and extinction – with durations on the scale of 100,000 yrs. Vrba (1993) argued that the pairing of origination with extinction during a turnover pulse is especially informative, and that it indicated that these two outcomes are part of a single continuum of species' responses to rapid habitat fragmentation and displacement: species undergo vicariance as metapopulations become disconnected, local populations shrink and either respond creatively to their shifting habitat through speciation or continue to decline into extinction. The TPH was originally developed as an explanation for the observed patterns of change in mammalian faunas across Africa during the late Neogene. A number of recent studies, however, suggest that similar dynamics affect marine plankton (e.g., Bendif *et al.*, 2019; Filatov *et al.*, 2021; Beaufort *et al.*, 2022; Bendif *et al.*, 2023).

The model of the Hirnantian mass extinction articulated by Brenchley *et al.* (2001, 2003) reflects the general conception of mass extinction as a two-step process in which an intense interval of species losses occurs in response to some catastrophic forcing that persists through a depauperate interregnum of some duration that is followed ultimately by a return to "normal" conditions accompanied by evolutionary radiation and ecological recovery. We argued earlier that our results indicate that this model is poorly fit by the graptolite record during the end

Ordovician events. The principal features of graptoloid species turnover during the LOME instead appear to correspond closely with those that Vrba identified as the likely drivers of a macroevolutionary turnover pulse. These include the coincidence of high rates of origination and extinction and their correspondence with times of strong habitat disruption caused by climate change and paced by Milankovitch cyclicity. Our data lack sufficient geographic resolution to demonstrate vicariance; however, the documented changes in metapopulation size do match predictions. The present analysis also generally lacks sufficient temporal resolution to demonstrate cyclicity in rates on the scale of 100 kyr throughout the Hirnantian, however, durations in the B3 bin set (equal sample size binning) during the interval of LOME-2 and LOME-3 are 100 kyr and the strong peaks in origination and extinction are fully captured (Fig. 6 A,B) and the $\hat{p}$ and $\hat{q}$ values for the B3 LOME-2 peak are the highest values obtained in any of our analyses. It is also noteworthy that the rapid changes in redox indicators at several sites in South China (e.g., Kozik *et al.*, 2022b; Jin Si-ding *et al.*, 2024; Wu Shuai-cai *et al.*, 2024) and by $\delta^{13}C$ and $\delta^{18}O$ values at Anticosti Island, Laurentia (Mauviel *et al.*, 2020; Zimmt *et al.*, 2024), among others, likely reflect astronomical forcing of climate variation. Such rapid fluctuations in redox and nutrient cycling undoubtedly added to the stress experienced by species during the LOME. Furthermore, that variability is likely to have empowered a selective ratchet that filtered organisms based on their capacity to accommodate the ensuing changes in habitats (size, location, and associated ecology). Dynesius and Jansson (2000) referred to this phenomenon as "orbitally-forced range dynamics" or ORD. It is likely that the Neograptina, which were already adapted to the higher levels of climate variability present in temperate zone regions, reacted to the Hirnantian ORD inherently differently than the paleotropical Diplograptina and that this difference contributed to the observed clade-level selectivity of the graptolite species turnover: extinction in one and speciation in the other.

Further analysis of species' biotope affiliation and geographic distribution through the LOME may permit testing of these ideas. Additional work is also needed to determine whether the observed replacement of diplograptines by neograptines was an expression of macroecological displacement driven by shared, limited resources or was simply a parallel but opposite response to icehouse conditions in the paleotropics.

4.8 Levels of Selection

Taken together, the principal features of graptolite species turnover during the LOME indicate that the late Katian oceanographic conditions produced by declining global temperature and the growth of the Gondwanan ice cap became increasingly inhospitable to the formerly dominant (wide-spread, abundant and long-lived) Diplograptina as their favored habitats shrank, were displaced offshore and became more temporally unstable. These changes were probably accompanied by increased seasonal upwelling and regionally extensive, shallow anoxia (in contrast to the more enduring stratified systems and deeper, more stable oxygen-depleted deep-water layers of the preceding Boda warm interval) – changes that also altered patterns of carbon and nutrient cycling along with phytoplankton community compositions. This forced diplograptine species into smaller habitable areas and probably led to smaller and less well-connected metapopulations. It is reasonable to infer that the resulting ORD increased the difficulty of tracking their diminished realized niche and maintaining viable populations. In contrast, the invading neograptine species, having formerly occupied temperate to subpolar sites, were exaptively well suited to the new prevailing oceanographic conditions of the latest Katian and Hirnantian oceans. Neograptines appear to have taken advantage of the cool, well-oxygenated shelf settings and seasonal upwelling zones that were increasingly dominated by green algae. These features combined with Milankovitch-driven oscillations in oceanographic conditions and the resulting orbitally forced range dynamics of graptolite populations likely facilitated vicariance and speciation among the Neograptina in step with increased extinction among the Diplograptina. In short, Late Ordovician climate change actuated a ratchet-like selection regime that ultimately drove the entire diplograptine clade into extinction while simultaneously advancing the species diversity of the Neograptina. This level of selectivity appears to be a particularly severe version of what now seems to be a common feature of mass extinctions (Jablonski 2005, 2017b, a).

The observed clade-level selectivity of the LOME turnover indicates that the forces driving this turnover acted upon species as selective-individuals. Although the differential extinction and origination of diplograptine and neograptine species surely involved the differential birth and death of individual graptolite colonies, the defining feature of this turnover event is that none of the diplograptine species possessed a sufficient reservoir of suitable variance to adapt to and survive the Hirnantian icehouse. The survival of *Paraorthograptus kimi* into LOME-3 near the end of the Persculptus Chron past the demise of its clade-mates might be regarded as an exception to that generality; however, this 'dead-clade walking' (sensu Jablonski 2002) merely proves the rule: species' survival without descendants is a macroevolutionary death. Conversely, several neograptine species lineages evidently possessed apt properties (e.g., niche properties related to temperature, resource utilization, modes of reproduction, etc.) that enabled many of their species not only to survive but also to undergo speciation at rates that routinely exceeded extinctions through the Hirnantian icehouse interval and its aftermath and to undergo the structural innovations that seeded the unique graptolite lineages of the Silurian Period (Melchin *et al.* 2011). Leaving aside the question of whether speciation and extinction are emergent, species-level traits or upward-caused effects of individual or population-level properties (see, for instance, Vrba & Gould 1986, Myers & Saupe 2013, Jablonski 2017b), it is clear that the lack of variance for these key macroevolutionary traits among the diplograptines and its residence exclusively among neograptines means that there cannot have been selection for differential survivorship and proliferation *within* these species (see Lloyd & Gould 1993, Gould & Lloyd 1999). Instead, selection can only have acted relative to variance in these traits *among* species, presumably as a consequence of their clade-specific apomorphies – that is, strictly via species selection. Indeed, selectivity of extinction and origination probabilities among clades (especially during intervals of intense selection as in the LOME) may be one of the most fruitful sources of direct evidence of species selection available in the fossil record.

5 Conclusions

Graptolite turnover from the latest Katian into the earliest Silurian exhibited an extraordinary level of clade selectivity. The Neograptina experienced high rates of speciation simultaneously with the high rates of extinction among diplograptine species; extinctions that ultimately led to the complete extirpation of the Diplograptina. The proliferation of the Neograptina appears to have resulted in some significant measure as an exaptive response to Hirnantian icehouse conditions, which created favorable circumstances in the paleotropics for the immigration of several temperate neograptine species. Their subsequent macroevolutionary radiation led to the origination of a series of novel graptolite clades including the uni-biserial Dimorphograpidae, the reticulated Retiolitinae and the wildly diverse Monograptidae (Melchin *et al.* 2011, Sadler *et al.* 2011, Bapst *et al.* 2012). Thus, although the LOME is often regarded as having exerted little selective effect on either the clade composition or ecological structure of marine communities (e.g., McGhee Jr *et al.* 2012, Krug & Patzkowsky 2015, Bush *et al.* 2020), planktic graptolites experienced the episode quite differently. This, in turn, suggests that the graptolite response to the LOME may provide a particularly sensitive indicator of the causes of that event.

Diplograptine and neograptine species occur together in the great preponderance of latest Katian and early Hirnantian samples and so they must have occupied more or less the same sites and water masses most of that time. This fact considered together with the trajectory of graptolite species extinction and origination during the LOME events and the strong clade-level selectivity of that turnover, indicate that oceanic anoxia was not a primary driver of species extinction among the zooplankton during the LOME. Instead, the widespread anoxia appears to have been a symptom of the altered oceanographic and climatic conditions associated with the Hirnantian glaciation, and in particular, reflected changes in oceanic circulation, nutrient supply, and phytoplankton productivity. We suggest that it was primarily these features, in combination with changes in habitat area and the displacement of habitats offshore, that drove graptolite species turnover during the LOME. Similarly, correlations between species turnover more generally and local facies and geochemical proxies may reflect common causes in which each was responding to the underlying paleoclimatic and oceanographic disruptions created during greenhouse–icehouse transitions. The long-term, multiphased history of turnover among graptolites through the Late Ordovician, with its coordinated rise and fall of origination and extinction rates in synchrony with climate-driven habitat change suggests that the LOME was not a unitary or even two-phased mass extinction at all but an extended episode of multiple successive turnover pulses (*a la* Vrba 1985, 1993) paced by orbital periodicity during the relatively short-lived but intense late Katian–Hirnantian glaciation and the following greenhouse.

6 Afterthought: The Late Ordovician Timescale

Zhang *et al.* (2025) present several new U-Pb radiometric dates derived from zircons sampled from ash beds in three sections in South China, including the Wangjiawan Hirnantian GSSP section. Based on these age results, Zhang *et al.* conclude that the Hirnantian Age lasted for only ~320 Kyr and that the *P. pacificus* Biozone interval had a duration of ~1 myr – much shorter than the intervals in our scaled composite (1.2 myr and 1.8 myr, respectively). These new dates indicate that the three glacial advance–retreat cycles (Mirus Subchron to early Persculptus Chron) that largely coincide with the LOME had durations of approximately 120 Kyr each, similar to those in the Quaternary glacial epoch. The Zhang *et al.* (2025) timescale also suggests that the duration of our B1–B3 bins in the Pacificus Chron interval are also about 120 kyr and those in the Mirus to Ascensus interval may have durations of only ~52 Kyr. These durations, in turn, suggest that the three major turnover pulses documented here correspond even more closely with those that Vrba (1985, 1993) identified in her model. The new dates also suggest, of course, that the LOME may have been much more intense than our present turnover rates reveal, with peak extinction and origination rates some four times higher than our rates and an increased contrast in rates between those in pre-Mirus Chron times and those in the main LOME turnover pulses. These revisions, however, do not alter the relative duration of three main turnover phases documented herein, which are not consistent with the claim in Zhang et al. (2025) that the LOME consisted of an extended, low-intensity phase in the early Hirnantian and a shorter, higher-rate turnover confined to the immediate postglacial interval. That apparent difference in rates through the course of the LOME again appears to be largely an artifact of the sampled record (as discussed earlier) and the general problem inherent in the comparison of rates measured over different interval lengths (see, for instance, Sadler 1981, Sheets & Mitchell 2001, Harmon *et al.* 2021).

References

Achab, A., Asselin, E., Desrochers, A., Riva, J. F., & Farley, C. 2011. Chitinozoan biostratigraphy of a new Upper Ordovician stratigraphic framework for Anticosti Island, Canada. *Geological Society of America Bulletin* 123(1–2):186–205.

Achab, A., Asselin, E., Desrochers, A., & Riva, J. F. 2013. The end-Ordovician chitinozoan zones of Anticosti Island, Québec: Definition and stratigraphic position. *Review of Palaeobotany and Palynology* 198:92–109.

Ahm, A.-S. C., Bjerrum, C. J., & Hammarlund, E. U. 2017. Disentangling the record of diagenesis, local redox conditions, and global seawater chemistry during the latest Ordovician glaciation. *Earth and Planetary Science Letters* 459:145–156.

Akaike, H. 1973. Information theory and an extension of the maximum likelihood principle. In B. N. Petrov and F. Csaki (eds.), *Second International Symposium on Information Theory*. Akademiai Kiado, Budapest, pp. 267–281.

Algeo, T. J., & Shen Jun. 2023. Theory and classification of mass extinction causation. *National Science Review* 11(1): 1–21.

Armstrong, H. A., & Coe, A. L. 1997. Deep-sea sediments record the geophysiology of the Late Ordovician glaciation. *Journal of the Geological Society of London* 154:929–934.

Bapst, D. W., Bullock, P. C., Melchin, M. J., Sheets, H. D., & Mitchell, C. E. 2012. Graptoloid diversity and disparity became decoupled during the Ordovician mass extinction. *Proceedings of the National Academy of Sciences* 109(9):3428–3433.

Barnosky, A. D. 2001. Distinguishing the effects of the Red queen and Court Jester on Miocene mammal evolution in the northern Rocky Mountains. *Journal of Vertebrate Paleontology* 21(1):172–185.

Bartlett, R., Elrick, M., Wheeley, J. R., *et al.* 2018. Abrupt global-ocean anoxia during the Late Ordovician–early Silurian detected using uranium isotopes of marine carbonates. *Proceedings of the National Academy of Sciences* 115(23):5896–5901.

Beaufort, L., Bolton, C. T., Sarr, A.-C., *et al.* 2022. Cyclic evolution of phytoplankton forced by changes in tropical seasonality. *Nature* 601 (7891):79–84.

Bendif, E. M., Nevado, B., Wong, E. L. Y., *et al.* 2019. Repeated species radiations in the recent evolution of the key marine phytoplankton lineage *Gephyrocapsa*. *Nature Communications* 10(1):4234.

Bendif, E. M., Probert, I., Archontikis, O. A., *et al.* 2023. Rapid diversification underlying the global dominance of a cosmopolitan phytoplankton. *The ISME Journal* 17(4):630–640.

Bergström, S. M., Chen Xu, Gutiérrez-Marco, J. C., & Dronov, A. 2009. The new chronostratigraphic classification of the Ordovician System and its relations to major regional series and stages and to $\delta 13C$ chemostratigraphy. *Lethaia* 42(1):97–107.

Bond, D. P., & Grasby, S. E. 2017. On the causes of mass extinctions. *Palaeogeography, Palaeoclimatology, Palaeoecology* 478:3–29.

Bond, D. P. G., & Grasby, S. E. 2020. Late Ordovician mass extinction caused by volcanism, warming, and anoxia, not cooling and glaciation. *Geology* 48(8):777–781.

Boyle, J., Sheets, H. D., Wu, S.-Y., *et al.* 2017. The impact of geographic range, sampling, ecology, and time on extinction risk in the volatile clade Graptoloida. *Paleobiology* 43(1):85–113.

Brenchley, P., Carden, G., Hints, L., *et al.* 2003. High-resolution stable isotope stratigraphy of Upper Ordovician sequences: Constraints on the timing of bioevents and environmental changes associated with mass extinction and glaciation. *Geological Society of America Bulletin* 115(1):89–104.

Brenchley, P. J., Marshall, J. D., Carden, G. A. F., *et al.* 1994. Bathymetric and isotopic evidence for a short-lived Late Ordovician glaciation in a greenhouse period. *Geology* 22:295–298.

Brenchley, P. J., Marshall, J. D., & Underwood, C. J. 2001. Do all mass extinctions represent an ecological crisis? Evidence from the Late Ordovician. *Geological Journal* 36(3–4):329–340.

Brett, C. E., Aucoin, C. D., Dattilo, B. F., *et al.* 2020. Revised sequence stratigraphy of the upper Katian Stage (Cincinnatian) strata in the Cincinnati Arch reference area: Geological and paleontological implications. *Palaeogeography, Palaeoclimatology, Palaeoecology* 540:109483.

Burnham, K. P., & Anderson, D. R. 1998. *Model Selection and Inference: A Practical Information-Theoretic Approach*. Springer, New York, p. 353.

Bush, A. M., Wang, S. C., Payne, J. L., & Heim, N. A. 2020. A framework for the integrated analysis of the magnitude, selectivity, and biotic effects of extinction and origination. *Paleobiology* 46(1):1–22.

Chen Qing, Fan Jun-xuan, Melchin, M. J., & Zhang Lin-na. 2014. Temporal and spatial distribution of the Wufeng Formation black shales (Upper Ordovician) in South China. *GFF* 136(1):55–59.

Chen Qing, Chen Ji-tao, Li Wen-jie, & Shi Zhen-sheng. 2023. Paleogeography and Paleoenvironment Across the Ordovician–Silurian Transition in the Yangtze Region. In Chen Xu, Wang Hong-yan, & Goldman, D., eds., *Latest Ordovician* to Early Silurian Shale Gas Strata of the Yangtze Region, China. Hangzhou: Springer and the Zhejiang University Press, pp. 151–181.

Chen Xu, Rong Jia-yu, Mitchell, C. E., et al. 2000. Late Ordovician to earliest Silurian graptolite and brachiopod biozonation from the Yangtze region, South China with a global correlation. *Geological Magazine* 137:623–650.

Chen Xu, Melchin, M. J., Fan Jun-xuan, & Mitchell, C. E. 2003. The Ashgillian graptolite fauna of the Yangtze regions and biogeographical distribution of diversity in the latest Ordovician. *Bulletin de la Societé géologique de France* 175:141–148.

Chen Xu, Rong Jia-yu, Li Yue, & Boucot, A. J. 2004. Facies patterns and geography of the Yangtze region, South China, through the Ordovician and Silurian transition. *Palaeogeography, Palaeoclimatology, Palaeoecology* 204(3):353–372.

Chen Xu, Fan Jun-xuan, Melchin, M. J., & Mitchell, C. E. 2005a. Hirnantian (Latest Ordovician) Graptolites from the Upper Yantze Region, China. *Palaeontology* 48(2):235–280.

Chen Xu, Melchin, M. J., Sheets, H. D., Mitchell, C. E., & Fan Jun-xuan. 2005b. Patterns and processes of latest Ordovician graptolite mass extinction and recovery based on data from South China. *Journal of Paleontology* 79:842–861.

Chen Xu, Rong Jia-yu, Fan Jun-xuan, et al. 2006a. The global boundary stratotype section and point (GSSP) for the base of the Hirnantian Stage (the uppermost of the Ordovician System). *Episodes* 29(3):183–197.

Chen Xu, Zhang Yuan-dong, & Fan Jun-xuan. 2006b. Ordovician graptolite evolutionary radiation: A review. *Geological Journal* 41(3–4):289–301.

Chen Yan, Cai Chun-fang, Qiu Zhen, & Lin Wei. 2021. Evolution of nitrogen cycling and primary productivity in the tropics during the Late Ordovician mass extinction. *Chemical Geology* 559:119926.

Connolly, S. R., & Miller, A. I. 2001. Joint estimation of sampling and turnover rates from fossil databases: Capture-mark-recapture methods revisited. *Paleobiology* 27:751–767.

Cooch, E., & White, G. C., eds. 2019. Program MARK, "A gentle introduction." www.phidot.org/software/mark/docs/book/.

Cooper, R. A., & Sadler, P. 2010. Facies preference predicts extinction risk in Ordovician graptolites. *Paleobiology* 32(2):167–187.

Cooper, R. A., Rigby, S., Loydell, D. K., & Bates, D. E. B. 2012. Palaeoecology of the Graptoloidea. *Earth-Science Reviews* 112(1–2):23–41.

Cooper, R. A., & Sadler, P. M. 2012. The Ordovician Period. In F. M. Gradstein, J. G. Ogg, M. D. Schmitz, and G. M. Ogg, eds. *The Geological Timescale 2012*. Elsevier B.V., Amsterdam, pp. 489–523.

Cooper, R. A., Sadler, P. M., Munnecke, A., & Crampton, J. S. 2014. Graptoloid evolutionary rates track Ordovician–Silurian global climate change. *Geological Magazine* 151(2):349–364.

Crampton, J. S., Cooper, R. A., Sadler, P. M., & Foote, M. 2016. Greenhouse–icehouse transition in the Late Ordovician marks a step change in extinction regime in the marine plankton. *Proceedings of the National Academy of Sciences* 113(6):1498–1503.

Crampton, J. S., Meyers, S. R., Cooper, R. A., et al. 2018. Pacing of Paleozoic macroevolutionary rates by Milankovitch grand cycles. *Proceedings of the National Academy of Sciences* 115(22):5686–5691.

Dahl, T. W., & Arens, S. K. M. 2020. The impacts of land plant evolution on Earth's climate and oxygenation state – An interdisciplinary review. *Chemical Geology* 547:119665.

Dahl, T. W., Hammarlund, E. U., Rasmussen, C. M. Ø., Bond, D. P. G., & Canfield, D. E. 2021. Sulfidic anoxia in the oceans during the Late Ordovician mass extinctions – insights from molybdenum and uranium isotopic global redox proxies. *Earth-Science Reviews* 220:103748.

Delabroye, A., Munnecke, A., Vecoli, M., et al. 2011. Phytoplankton dynamics across the Ordovician/Silurian boundary at low palaeolatitudes: Correlations with carbon isotopic and glacial events. *Palaeogeography, Palaeoclimatology, Palaeoecology* 312(1–2): 79–97.

Deng Yi-ing, Fan Jun-xuan, Zhang Shu-han, et al. 2021. Timing and patterns of the Great Ordovician Biodiversification Event and Late Ordovician mass extinction: Perspectives from South China. *Earth-Science Reviews* 220:103743.

Denis, M., Buoncristiani, J.-F., Konaté, M., Ghienne, J.-F., & Guiraud, M. 2007. Hirnantian glacial and deglacial record in SW Djado Basin (NE Niger). *Geodinamica acta* 20(3):177–195.

Du Xue-bin, Lu Yong-chao, Duan Dan, et al. 2020. Was volcanic activity during the Ordovician–Silurian transition in South China part of a global phenomenon? Constraints from zircon U–Pb dating of volcanic ash beds in black shales. *Marine and Petroleum Geology* 114:104209.

Dynesius, M., & Jansson, R. 2000. Evolutionary consequences of changes in species' geographical distributions driven by Milankovitch climate oscillations. *Proceedings of the National Academy of Sciences* 97(16):9115–9120.

Faith, J. T., & Behrensmeyer, A. K. 2013. Climate change and faunal turnover: testing the mechanics of the turnover-pulse hypothesis with South African fossil data. *Paleobiology* 39(4):609–627.

Fan Jun-xuan, & Chen Xu. 2007. Preliminary report on the Late Ordovician graptolite extinction in the Yangtze region. *Palaeogeography, Palaeoclimatology, Palaeoecology* 245(1):82–94.

Fan Jun-xuan, Shen Shu-zhong, Erwin, D. H., *et al.* 2020. A high-resolution summary of Cambrian to Early Triassic marine invertebrate biodiversity. *Science* 367(6475): 272.

Fanton, K. C., & Holmden, C. 2007. Sea-level forcing of carbon isotope excursions in epeiric seas: Implications for chemostratigraphy. *Canadian Journal of Earth Sciences* 44(6):807–818.

Filatov, D. A., Bendif, E. M., Archontikis, O. A., Hagino, K., & Rickaby, R. E. M. 2021. The mode of speciation during a recent radiation in open-ocean phytoplankton. *Current Biology* 31(24):5439–5449.e5.

Finnegan, S., Bergmann, K., Eiler, J. M., *et al.* 2011. The Magnitude and Duration of Late Ordovician–Early Silurian Glaciation. *Science* 331(6019): 903–906.

Finnegan, S., Fike, D. A., Jones, D., & Fischer, W. W. 2012a. A Temperature-Dependent Positive Feedback on the Magnitude of Carbon Isotope Excursions. *Geoscience Canada* 39(3):122–131.

Finnegan, S., Heim, N. A., Peters, S. E., & Fischer, W. W. 2012b. Climate change and the selective signature of the Late Ordovician mass extinction. *Proceedings of the National Academy of Sciences* 109(18):6829–6834.

Finney, S. C., Berry, W. B. N., Cooper, J. D., *et al.* 1999. Late Ordovician mass extinction: A new perspective from stratigraphic sections in central Nevada. *Geology* 27(3):215–218.

Finney, S. C. 2001. Species diversification during mass extinction: Graptolites in the Late Ordovician, pp. A-213. *Geological Society of America Annual Meeting. Abstracts with Programs*, Boston, MA.

Finney, S. C., Berry, W. B. N., & Cooper, J. D. 2007. The influence of denitrifying seawater on graptolite extinction and diversification during the Hirnantian (Latest Ordovician) mass extinction event. *Lethaia* 40:281–291.

Foote, M. 2000. Origination and extinction components of taxonomic diversity: General problems. In D. H. Erwin, and S. L. Wing, eds. *Deep Time: Paleobiology's Perspective*. The Paleontological Society, Lawrence, Kansas, pp. 74–102.

Foote, M. 2001. Inferring temporal patterns of preservation, origination, and extinction from taxonomic survivorship analysis. *Paleobiology* 27(4):602–630.

Foote, M. 2003. Origination and extinction through the Phanerozoic: A new approach. *The Journal of Geology* 111(2):125–148.

Foote, M., Sadler, P. M., Cooper, R. A., & Crampton, J. S. 2019. Completeness of the known graptoloid palaeontological record. *Journal of the Geological Society* 176(6), 1038–105.

Fortey, R. A., & Cooper, R. A. 1986. A phylogenetic classification of the graptoloids. *Palaeontology* 29(4):631–654.

Ghienne, J.-F., Desrochers, A., Vandenbroucke, T. R., et al. 2014. A Cenozoic-style scenario for the end-Ordovician glaciation. *Nature Communications* 5:4485.

Goldman, D., Mitchell, C. E., Melchin, M. J., et al. 2011. Biogeography and Mass Extinction: Extirpation and re-invasion of *Normalograptus* species (Graptolithina) in the Late Ordovician Palaeotropics. *Proceedings of the Yorkshire Geological Society* 58(4):227–246.

Goldman, D., Maletz, J., Melchin, M. J., & Fan Jun-xuan. 2014. Lower Palaeozoic Graptolite Biogeography. In D. A. T. Harper, and T. Servais, eds. Early Palaeozoic Palaeobiogeography and Palaeogeography. *Geological Society of London Memoir* 38, pp. 415–428.

Goldman, D., Sadler, P. M., Leslie, S. A., et al. 2020. Chapter 20: The Ordovician Period. In F. M. Gradstein, J. G. Ogg, M. D. Schmitz, and G. M. Ogg, eds. *Geologic Time Scale 2020*. Elsevier, Amsterdam, pp. 631–694.

Gorjan, P., Kaiho, K., Fike, D. A., & Chen Xu. 2012. Carbon- and sulfur-isotope geochemistry of the Hirnantian (Late Ordovician) Wangjiawan (Riverside) section, South China: Global correlation and environmental event interpretation. *Palaeogeography, Palaeoclimatology, Palaeoecology* 337–338:14–22.

Gould, S. J., & Lloyd, E. A. 1999. Individuality and adaptation across levels of selection: how shall we name and generalize the unit of Darwinism? *Proceedings of the National Academy of Sciences* 96(21):11904–11909.

Hammarlund, E. U., Dahl, T. W., Harper, D. A. T., et al. 2012. A sulfidic driver for the end-Ordovician mass extinction. *Earth and Planetary Science Letters* 331–332(May):128–139.

Harmon, L. J., Pennell, M. W., Henao-Diaz, L. F., et al. 2021. Causes and consequences of apparent timescaling across all estimated evolutionary rates. *Annual Review of Ecology, Evolution, and Systematics* 52:587–609.

Harper, D. A. T., Hammarlund, E. U., & Rasmussen, C. M. 2014. End Ordovician extinctions: A coincidence of causes. *Gondwana Research* 25(4):1294–1307.

Harper, D. A. T. 2023. Late Ordovician Mass Extinction: Earth, fire and ice. *National Science Review* 11(1):nwad319.

Holland, S. M., & Patzkowsky, M. E. 2015. The stratigraphy of mass extinction. *Palaeontology* 58(5):903–924.

Holland, S. M. 2016. Ecological disruption precedes mass extinction. *Proceedings of the National Academy of Sciences* 113(30):8349–8351.

Holland, S. M. 2020. The stratigraphy of mass extinctions and recoveries. *Annual Review of Earth and Planetary Sciences* 48:75–97.

Holland, S. M. 2023. The contrasting controls on the occurrence of fossils in marine and nonmarine systems. *Bollettino della Società Paleontologica Italiana* 62(1):1–25.

Holmden, C., Mitchell, C. E., LaPorte, D. F., *et al.* 2013. Nd isotope records of late Ordovician sea-level change: Implications for glaciation frequency and global stratigraphic correlation. *Palaeogeography, Palaeoclimatology, Palaeoecology* 386(18):131–144.

Hu Dong-ping, Li Meng-han, Zhang Xiao-lin, *et al.* 2020. Large mass-independent sulphur isotope anomalies link stratospheric volcanism to the Late Ordovician mass extinction. *Nature Communications* 11(1):2297.

Hu Dong-ping, Li Meng-han, Jiu-bin, C., *et al.* 2021. Major volcanic eruptions linked to the Late Ordovician mass extinction: Evidence from mercury enrichment and Hg isotopes. *Global and Planetary Change* 196:103374.

Hu Rui-ning, Tan Jing-qiang, Wang Wen-hui, *et al.* 2024. Coupling of nitrogen biogeochemical cycling and phytoplankton community structure before and after the Late Ordovician mass extinction in South China. *Chemical Geology* 647:121933.

Husinec, A., & Leslie, S. A. 2022. Continuous record of Upper Ordovician (Katian) to lower Silurian (Telychian) global δ^{13}C excursions in the Williston Basin. *Terra Nova* 34(4):314–322.

Immenhauser, A., Holmden, C., & Patterson, W. P. 2008. Interpreting the carbon-isotope record of ancient shallow epeiric seas: Lessons from the recent. In: B. R. Pratt and C. Holmden, eds., *Dynamics of Epeiric Seas*. Geological Society of Canada Special Publication 48, pp. 135–174.

Jablonski, D. 2002. Survival without recovery after mass extinctions. *Proceedings of the National Academy of Sciences* 99(12):8139–8144.

Jablonski, D. 2005. Mass extinctions and macroevolution. *Paleobiology* 31(S2):192–210.

Jablonski, D. 2017a. Approaches to macroevolution: 1. General concepts and origin of variation. *Evolutionary Biology* 44(4):427–450.

Jablonski, D. 2017b. Approaches to macroevolution: 2. Sorting of variation, some overarching issues, and general conclusions. *Evolutionary Biology* 44(4):451–475.

Jeon, J., Li, Y., Kershaw, S., *et al.* 2022. Nearshore warm-water biota development in the aftermath of the Late Ordovician Mass Extinction in South China. *Palaeogeography, Palaeoclimatology, Palaeoecology* 603:111182.

Jin, J., & Harper, D. A. T. 2024. An Edgewood-type Hirnantian fauna from the Mackenzie Mountains, northwestern margin of Laurentia. *Journal of Paleontology* 98(1):13–39.

Jin Si-ding, Deng Hu-cheng, Zhu Xing, *et al.* 2020. Orbital control on cyclical organic matter accumulation in Early Silurian Longmaxi Formation shales. *Geoscience Frontiers* 11(2):533–545.

Jin Si-ding, Cao Hai-yang, Hou Ming-cai, *et al.* 2024. Orbital and Millennial-Scale Cycles Through the Hirnantian (Late Ordovician) in Southern China. *Geochemistry, Geophysics, Geosystems* 25(1): e2023GC011127.

Jones, D. S., Martini, A. M., Fike, D. A., & Kaiho, K. 2017. A volcanic trigger for the Late Ordovician mass extinction? Mercury data from south China and Laurentia. *Geology* 45(7):631–634.

Jones, D. S., Brothers, R. W., Crüger Ahm, A.-S., *et al.* 2020. Sea level, carbonate mineralogy, and early diagenesis controlled $\delta^{13}C$ records in Upper Ordovician carbonates. *Geology* 48(2):194–199.

Kemple, W. G., Sadler, P. M., & Strauss, D. J. 1995. Extending graphic correlation to many dimensions: Stratigraphic correlation as constrained optimization. In K. O. Mann, and H. R. Lane, eds. *Graphic Correlation*. SEPM Society for Sedimentary Geology, Tulsa, Oklahoma, pp. 65–82.

Koren, T. N., Oradovskaya, M. M., Pylma, L. J., Sobolevskaya, R. F., & Chugaeva, M. N. 1983. The Ordovician and Silurian boundary in the northeast of the U.S.S.R. Nauka Publishers, Leningrad, p. 205.

Koren', T. N., & Sobolevskaya, R. F. 2008. The regional stratotype section and point for the base of the Hirnantian Stage (the uppermost Ordovician) at Mirny Creek, Omulev Mountains, Northeast Russia. *Estonian Journal of Earth Sciences* 57(1):1–10.

Kozik, N. P., Gill, B. C., Owens, J. D., Lyons, T. W., & Young, S. A. 2022a. Geochemical records reveal protracted and differential marine redox change associated with Late Ordovician climate and mass extinctions. *AGU Advances* 3(1):e2021AV000563.

Kozik, N. P., Young, S. A., Newby, S. M., *et al.* 2022b. Rapid marine oxygen variability: Driver of the Late Ordovician mass extinction. *Science Advances* 8(46):eabn8345.

Kraft, P., Štorch, P., & Mitchell, C. E. 2015. Graptolites of the Králův Dvůr Formation (mid Katian to earliest Hirnantian, Czech Republic). *Bulletin of Geosciences* 90(1):195–225.

Kröger, B., Franeck, F., & Rasmussen, C. M. 2019. The evolutionary dynamics of the early Palaeozoic marine biodiversity accumulation. *Proceedings of the Royal Society* B 286(1909):20191634.

Krug, A. Z., & Patzkowsky, M. E. 2015. Phylogenetic clustering of origination and extinction across the Late Ordovician mass extinction. *Plos One* 10(12): e0144354.

Kump, L. R., Arthur, M., Patzkowsky, M., *et al.* 1999. A weathering hypothesis for glaciation at high atmospheric pCO_2 during the Late Ordovician. *Palaeogeography, Palaeoclimatology, Palaeoecology* 152(1–2):173–187.

Kump, L. R., & Arthur, M. A. 1999. Interpreting carbon-isotope excursions: carbonates and organic matter. *Chemical Geology* 161(1–3):181–198.

LaPorte, D. F., Holmden, C., Patterson, W. P., *et al.* 2009. Local and global perspectives on carbon and nitrogen cycling during the Hirnantian glaciation. *Palaeogeography, Palaeoclimatology, Palaeoecology* 276(1–4):182–195.

Lenton, T. M., Crouch, M., Johnson, M., Pires, N., & Dolan, L. 2012. First plants cooled the Ordovician. *Nature Geoscience* 5(2):86–89.

Lenton, T. M., Dahl, T. W., Daines, S. J., *et al.* 2016. Earliest land plants created modern levels of atmospheric oxygen. *Proceedings of the National Academy of Sciences* 113(35):9704–9709.

Li Chao, Zhang Jun-peng, Li Wen-jie, *et al.* 2021. Multiple glacio-eustatic cycles and associated environmental changes through the Hirnantian (Late Ordovician) in South China. *Global and Planetary Change* 207:103668.

Li Na, Li Chao, Fan Jun-xuan, *et al.* 2019. Sulfate-controlled marine euxinia in the semi-restricted inner Yangtze Sea (South China) during the Ordovician-Silurian transition. *Palaeogeography, Palaeoclimatology, Palaeoecology* 534:109281.

Li Na, Li Chao, Algeo, T. J., *et al.* 2021. Redox changes in the outer Yangtze Sea (South China) through the Hirnantian Glaciation and their implications for the end-Ordovician biocrisis. *Earth-Science Reviews* 212:103443.

Liang Yu, Liu Zerui Ray, Algeo, T. J., *et al.* 2024. Contrasting dynamics of marine bacterial-algal communities between the two main pulses of the Late Ordovician Mass Extinction. *Earth and Planetary Science Letters* 645:118956.

Ling Ming-xing, Zhan Ren-bin, Wang Guang-xu, *et al.* 2019. An extremely brief end Ordovician mass extinction linked to abrupt onset of glaciation. *Solid Earth Sciences* 4(4):190–198.

Liow, L. H., & Nichols, J. D. 2010. Estimating Rates and Probabilities of Origination and Extinction Using Taxonomic Occurrence Data: Capture-Mark-Recapture (CMR) Approaches. *The Paleontological Society Papers* 16:81–94.

Liu Mu, Chen Dai-zhao, Jiang Lei, *et al.* 2022. Oceanic anoxia and extinction in the latest Ordovician. *Earth and Planetary Science Letters* 588:117553.

Liu Yu, Li Chao, Fan Jun-xuan, & Algeo, T. J. 2020. Elevated marine productivity triggered nitrogen limitation on the Yangtze Platform (South China) during the Ordovician-Silurian transition. *Palaeogeography, Palaeoclimatology, Palaeoecology* 554:109833.

Lloyd, E. A., & Gould, S. J. 1993. Species selection on variability. *Proceedings of the National Academy of Sciences* 90(2):595–599.

Longman, J., Mills, B. J. W., Manners, H. R., Gernon, T. M., & Palmer, M. R. 2021. Late Ordovician climate change and extinctions driven by elevated volcanic nutrient supply. *Nature Geoscience* 14(12):924–929.

Lu Yang-bo, Huang Chun-ju, Jiang Shu, *et al.* 2019. Cyclic late Katian through Hirnantian glacioeustasy and its control of the development of the organic-rich Wufeng and Longmaxi shales, South China. *Palaeogeography, Palaeoclimatology, Palaeoecology* 526:96–109.

Lu Yang-bo, Shen Jun, Wang Yu-xuan, *et al.* 2022. Seawater sources of Hg enrichment in Ordovician-Silurian boundary strata, South China. *Palaeogeography, Palaeoclimatology, Palaeoecology* 601:111156.

Ludvigson, G. A., Witzke, B. J., González, L. A., *et al.* 2004. Late Ordovician (Turinian-Chatfieldian) carbon isotope excursions and their stratigraphic and paleoceanographic significance. *Palaeogeography, Palaeoclimatology, Palaeoecology* 210(2–4):187–214.

Luo Gen-ming, Algeo, T. J., Zhan Ren-bin, *et al.* 2016. Perturbation of the marine nitrogen cycle during the Late Ordovician glaciation and mass extinction. *Palaeogeography, Palaeoclimatology, Palaeoecology* 448:339–348.

Luo Gen-ming, Yang Huan, Algeo, T. J., Hallmann, C., & Xie Shu-cheng. 2018. Lipid biomarkers for the reconstruction of deep-time environmental conditions. *Earth-Science Reviews* 189 (February 2019): 99–124.

Maletz, J. 2023. (coordinating author) *Treatise on Invertebrate Paleontology, Part V, Hemichordata, Second Revision, Including Enteropneusta, Pterobranchia (Graptolithina)*. The University of Kansas, Paleontological Institute, Lawrence, Kansas, p. 548.

Marshall, J. D., & Middleton, P. D. 1990. Changes in marine isotopic composition and the late Ordovician glaciation. *Journal of the Geological Society* 147(1):1–4.

Mauviel, A., & Desrochers, A. 2016. A high-resolution, continuous δ^{13}C record spanning the Ordovician–Silurian boundary on Anticosti Island, eastern Canada. *Canadian Journal of Earth Sciences* 53(8):795–801.

Mauviel, A., Sinnesael, M., & Desrochers, A. 2020. The stratigraphic and geochemical imprints of Late Ordovician glaciation on far-field neritic carbonates, Anticosti Island, eastern Canada. *Palaeogeography, Palaeoclimatology, Palaeoecology* 543:109579.

McGhee Jr, G. R., Sheehan, P. M., Bottjer, D. J., & Droser, M. L. 2012. Ecological ranking of Phanerozoic biodiversity crises: the Serpukhovian (early Carboniferous) crisis had a greater ecological impact than the end-Ordovician. *Geology* 40(2):147–150.

Melchin, M. J., & Mitchell, C. E. 1991. Late Ordovician extinction in the Graptoloidea. In C. R. Barnes, and S. H. Williams, eds. *Advances in Ordovician Geology*. Geological Survey of Canada Paper 90–9, Ottawa, pp. 143–156.

Melchin, M. J. 1998. Morphology and phylogeny of some early Silurian 'diplograptid' genera from Cornwallis Island, Arctic Canada. *Palaeontology* 41(2):263–315, 7 pls.

Melchin, M. J., Holmden, C., & Williams, S. H. 2003. Correlation of graptolite biozones, chitinozoan biozones, and carbon isotope curves through the Hirnantian. In G. L. Albanesi, M. S. Beresi, and S. H. Peralta, eds. *Ordovician from the Andes*. INSUEGO, Serie Correlación Geológica, Comunicarte Editorial, Tucumán, Argentina, pp. 101–104.

Melchin, M. J., & Holmden, C. 2006. Carbon isotope chemostratigraphy in Arctic Canada: Sea-level forcing of carbonate platform weathering and implications for Hirnantian global correlation. *Palaeogeography, Palaeoclimatology, Palaeoecology* 234(186–200).

Melchin, M. J. 2008. Restudy of some Ordovician–Silurian boundary graptolites from Anticosti Island, Canada, and their biostratigraphic significance. *Lethaia* 41(2):155–162.

Melchin, M. J., Mitchell, C. E., Naczk-Cameron, A., Fan Junxuan, & Loxton, J. 2011. Phylogeny and Adaptive Radiation of the Neograptina (Graptoloida) During the Hirnantian Mass Extinction and Silurian Recovery. *Proceedings of the Yorkshire Geological Society* 58(4):281–309.

Melchin, M. J., Sadler, P. M., Cramer, B. D., et al. 2012. Chapter 21: The Silurian Period, pp. 525–558 *in* F. M. Gradstein, J. G. Ogg, M. Schmitz, and G. M. Ogg, eds. *The Geological Timescale 2012*. Elsevier, Amsterdam.

Melchin, M. J., Mitchell, C. E., Holmden, C., & Štorch, P. 2013. Environmental changes in the Late Ordovician-early Silurian: Review and new insights from black shales and nitrogen isotopes. *Geological Society of America Bulletin* 125(11/12):1635–1670.

Melchin, M. J., Sheets, H. D., Mitchell, C. E., & Fan, J. 2017. A new approach to quantifying stratigraphical resolution: application to global stratotypes. *Lethaia* 50(3):407–423.

Mitchell, C. E. 1987. Evolution and phylogenetic classification of the Diplograptacea. *Palaeontology* 30(2):353–405.

Mitchell, C. E. 1990. Directional macroevolution of the diplograptacean graptolites: a product of astogenetic heterochrony and directed speciation. In P. D. Taylor, and G. P. Larwood, eds. *Major Evolutionary Radiations*. Clarendon Press, Oxford, pp. 235–264.

Mitchell, C. E., Goldman, D., Klosterman, S. L., et al. 2007a. Phylogeny of the Ordovician Diplograptoidea. *Acta Palaeontologia Sinica* 46 (supplement):332–339.

Mitchell, C. E., Sheets, H. D., Belscher, K., et al. 2007b. Species abundance changes during mass extinction and the inverse Signor-Lipps Effect: Apparently abrupt graptolite mass extinction as an artifact of sampling. *Acta Palaeontologica Sinica* 46 (supplement):340–346.

Mitchell, C. E., Maletz, J., & Goldman, D. 2009. What is Diplograptus? *Bulletin of Geosciences* 84(1):27–34.

Mitchell, C. E., Štorch, P., Holmden, C., Melchin, M. J., & Gutierrez-Marco, J. C. 2011. New stable isotope data and fossils from the Hirnantian Stage in Bohemia and Spain: implications for correlation and paleoclimate. In J. C. Gutierrez-Marco, I. Rábano, and D. García-Bellido, eds. Ordovician of the World. *Cuadernos del Museo Geominero*, 14. Instituto Geológico y Minero de España, Madrid, pp. 371–378.

Mitchell, C. E., & Melchin, M. J. 2020. Late Ordovician mass extinction caused by volcanism, warming, and anoxia, not cooling and glaciation: COMMENT. *Geology* 48(8):e509-e509.

Müller, J., Joachimski, M. M., Lehnert, O., Männik, P., & Sun, Y. 2024. Phosphorus cycling during the Hirnantian glaciation. *Palaeogeography, Palaeoclimatology, Palaeoecology* 634:111906.

Myers, C. E., & Saupe, E. E. 2013. A macroevolutionary expansion of the modern synthesis and the importance of extrinsic abiotic factors. *Palaeontology* 56(6):1179–1198.

Olivarez Lyle, A., & Lyle, M. W. 2006. Missing organic carbon in Eocene marine sediments: Is metabolism the biological feedback that maintains end-member climates? *Paleoceanography* 21(2).

Parnell, J., & Foster, S. 2012. Ordovician ash geochemistry and the establishment of land plants. *Geochemical Transactions* 13(1):1–7.

Pohl, A., Lu, Z., Lu, W., et al. 2021. Vertical decoupling in Late Ordovician anoxia due to reorganization of ocean circulation. *Nature Geoscience* 14(11):868–873.

Porada, P., Lenton, T. M., Pohl, A., et al. 2016. High potential for weathering and climate effects of non-vascular vegetation in the Late Ordovician. *Nature Communications* 7:12113.

Qiu Zhen, Zou Cain-eng, Mills, B. J. W., et al. 2022. A nutrient control on expanded anoxia and global cooling during the Late Ordovician mass extinction. *Communications Earth & Environment* 3(1):82.

Rasmussen, C. M. Ø., Kröger, B., Nielsen, M. L., & Colmenar, J. 2019. Cascading trend of Early Paleozoic marine radiations paused by Late

Ordovician extinctions. *Proceedings of the National Academy of Sciences* 116(15):7207.

Rohrssen, M., Love, G. D., Fischer, W., Finnegan, S., & Fike, D. A. 2013. Lipid biomarkers record fundamental changes in the microbial community structure of tropical seas during the Late Ordovician Hirnantian glaciation. *Geology* 41(2):127–130.

Rong Jiayu, Harper, D. A. T., Huang Bing, et al. 2020. The latest Ordovician Hirnantian brachiopod faunas: New global insights. *Earth-Science Reviews* 208:103280.

Sadler, P. M. 1981. Sediment accumulation rates and the completeness of stratigraphic sections. *Journal of Geology* 89:569–584.

Sadler, P. M., & Kemple, W. G. 1995. Using rapid multidimensional, graphic correlation to evaluate chronostratigraphic models for the Mid-Ordovician of the Mohawk valley, New York. In J. D. Cooper, M. L. Droser, and S. C. Finney, eds. *Ordovician Odyssey: Short Papers for the Seventh International Symposium on the Ordovician System*. The Pacific Section Society for Sedimentary Geology (SEPM), Fullerton, California, pp. 257–260.

Sadler, P. M., Kemple, W. G., & Kooser, M. A. 2003. Chapter 13. Contents of the compact disk – CONOP9 programs for solving the stratigraphic correlation and seriation problems as constrained optimization. In P. J. Harries, ed. *High resolution approaches in stratigraphic paleontology.* Kluwer Academic Publishers, Dordrecht, pp. 461–65.

Sadler, P. M., Cooper, R. A., & Melchin, M. J. 2011. Sequencing the graptoloid clade: building a global diversity curve from local range charts, regional composites and global time-lines. *Proceedings of the Yorkshire Geological Society* 58(4):329–343.

Saltzman, M. R. 2005. Phosphorus, nitrogen, and the redox evolution of the Paleozoic oceans. *Geology* 33(7):573–576.

Saltzman, M. R., & Young, S. A. 2005. Long-lived glaciation in the Late Ordovician? Isotopic and sequence-stratigraphic evidence from western Laurentia. *Geology* 33(2):109–112.

Sheehan, P. M. 2001. The Late Ordovician Mass Extinction. *Annual Review of Earth and Planetary Sciences* 29:331–364.

Sheets, H. D., & Mitchell, C. E. 2001. Uncorrelated change produces the apparent dependence of evolution rate on interval. *Paleobiology* 27(3):429–445.

Sheets, H. D., Mitchell, C. E., Izard, Z. T., et al. 2012. Horizon annealing: a collection-based approach to automated sequencing of the fossil record. *Lethaia* 45(4):532–547.

Sheets, H. D., Mitchell, C. E., Melchin, M. J., *et al.* 2016. Graptolite community responses to global climate change and the Late Ordovician mass extinction. *Proceedings of the National Academy of Sciences* 113(30):8380–8385.

Shen Jia-heng, Pearson, A., Henkes, G. A., *et al.* 2018. Improved efficiency of the biological pump as a trigger for the Late Ordovician glaciation. *Nature Geoscience* 11(7):510–514.

Shen Jun, Algeo, T. J., Chen Jiu-bin, *et al.* 2019. Mercury in marine Ordovician/Silurian boundary sections of South China is sulfide-hosted and non-volcanic in origin. *Earth and Planetary Science Letters* 511:130–140.

Shen Jun, Algeo, T. J., & Feng Qing-lai. 2022. Mercury isotope evidence for a non-volcanic origin of Hg spikes at the Ordovician-Silurian boundary, South China. *Earth and Planetary Science Letters* 594:117705.

Sinnesael, M., McLaughlin, P. I., Desrochers, A., *et al.* 2021. Precession-driven climate cycles and time scale prior to the Hirnantian glacial maximum. *Geology* 49(11):1295–1300.

Smolarek, J., Marynowski, L., Trela, W., Kujawski, P., & Simoneit, B. R. T. 2017. Redox conditions and marine microbial community changes during the end-Ordovician mass extinction event. *Global and Planetary Change* 149:105–122.

Smolarek-Lach, J., Marynowski, L., Trela, W., & Wignall, P. B. 2019. Mercury spikes indicate a volcanic trigger for the Late Ordovician mass extinction event: An example from a deep shelf of the peri-Baltic region. *Scientific Reports* 9(1):3139.

Song Li-cai, Chen Qing, Li Hui-jun, & Deng Chang-zhou. 2023. Roller-coaster atmospheric-terrestrial-oceanic-climatic system during Ordovician-Silurian transition: Consequences of large igneous provinces. *Geoscience Frontiers* 14(3):101537.

Stanley, S. M. 2010. Relation of Phanerozoic stable isotope excursions to climate, bacterial metabolism, and major extinctions. *Proceedings of the National Academy of Sciences* 107(45):19185–19189.

Štorch, P., Mitchell, C. E., Finney, S. C., & Melchin, M. J. 2011. Uppermost Ordovician (upper Katian-Hirnantian) graptolites of north-central Nevada, U.S.A. *Bulletin of Geosciences* 86(2):301–386.

Štorch, P. 2023. Graptolite biostratigraphy and biodiversity dynamics in the Silurian System of the Prague Synform (Barrandian area, Czech Republic). *Bulletin of Geosciences* 98(1):1–78.

Vandenbroucke, T. R., Emsbo, P., Munnecke, A., *et al.* 2015. Metal-induced malformations in early Palaeozoic plankton are harbingers of mass extinction. *Nature Communications* 6:7966.

Vrba, E. S. 1985. Environment and evolution: Alternative causes of the temporal distribution of evolutionary events. *South African Journal of Science* 81:229–236.

Vrba, E. S. 1993. Turnover-pulses, the Red Queen, and related topics. *American Journal of Science* 293(A):418–452.

Vrba, E. S., & Gould, S. J. 1986. The hierarchical expansion of sorting and selection: sorting and selection cannot be equated. *Paleobiology* 12(2):217–228.

Wang Guang-xu, Zhan Ren-bin, & Percival, I. G. 2019. The end-Ordovician mass extinction: A single-pulse event? *Earth-Science Reviews* 192:15–33.

Wang, S. C., & Zhong, L. 2018. Estimating the number of pulses in a mass extinction. *Paleobiology* 44(2):199–218.

Williams, S. H. 1982. The late Ordovician graptolite fauna of the Anceps bands at Dob`s Linn, southern Scotland. *Geologica et Palaeontologica* 16:29–56.

Williams, S. H. 1983. The Ordovician-Silurian boundary graptolite fauna of Dob`s Linn, southern Scotland. *Palaeontology* 26(3):605–639.

Wu Shuai-cai, Chen Lei, Xiong Min, et al. 2024. Depositional conditions of shale lithofacies during the Late Ordovician–Early Silurian in the Upper Yangtze area, SW China: Responses to sea-level changes. *Marine and Petroleum Geology* 161:106696.

Yan De-tain, Wang Hua, Fu Qi-long, et al. 2015. Organic matter accumulation of Late Ordovician sediments in North Guizhou Province, China: Sulfur isotope and trace element evidences. *Marine and Petroleum Geology* 59:348–358.

Yang Sheng-chao, Hu Wen-xuan, Wang Xiao-lin, et al. 2019. Duration, evolution, and implications of volcanic activity across the Ordovician–Silurian transition in the Lower Yangtze region, South China. *Earth and Planetary Science Letters* 518:13–25.

Yang Sheng-chao, Hu Wen-xuan, Wang Xiao-lin, & Fan Jun-xuan. 2021. Nitrogen isotope evidence for a redox-stratified ocean and eustasy-driven environmental evolution during the Ordovician–Silurian transition. *Global and Planetary Change* 207:103682.

Yang Sheng-chao, Fan Jun-xuan, Algeo, T. J., et al. 2024. Steep oceanic DIC $\delta^{13}C$ depth gradient during the Hirnantian Glaciation. *Earth-Science Reviews* 255:104840.

Yang Sheng-chao, & Fan Jun-xuan. 2025. Decreased marine organic carbon burial during the Hirnantian glaciation. *Earth and Planetary Science Letters* 654:119240.

Yang Xiang-rong, Yan De-tian, Li Tong, et al. 2020. Oceanic environment changes caused the Late Ordovician extinction: evidence from geochemical and Nd isotopic composition in the Yangtze area, South China. *Geological Magazine* 157(4):651–665.

Yang Xiang-rong, Yan De-tian, Chen Dai-zhao, *et al.* 2021. Spatiotemporal variations of sedimentary carbon and nitrogen isotopic compositions in the Yangtze Shelf Sea across the Ordovician-Silurian boundary. *Palaeogeography, Palaeoclimatology, Palaeoecology* 567:110257.

Young, S. A., Benayoun, E., Kozik, N. P., *et al.* 2020. Marine redox variability from Baltica during extinction events in the latest Ordovician–early Silurian. *Palaeogeography, Palaeoclimatology, Palaeoecology* 554:109792.

Young, S. A., Edwards, C. T., Ainsaar, L., Lindskog, A., & Saltzman, M. R. 2023. Seawater signatures of Ordovician climate and environment. *Geological Society, London, Special Publications* 532(1):137–156.

Zan Bo-wen, Mou Chuan-long, Lash, G. G., *et al.* 2024. Upwelling-driven biogenic silica accumulation in the Yangtze Sea, South China during late Ordovician to early Silurian time: A possible link with the global climatic transitions. *Sedimentary Geology*: 106571.

Zhang Jun-peng, Li Chao, Zhong Yang-yang, *et al.* 2024. Linking carbon cycle perturbations to the Late Ordovician glaciation and mass extinction: A modeling approach. *Earth and Planetary Science Letters* 631:118635.

Zhang, Z., Yang, C., Sahy, D., *et al.* 2025. Tempo of the Late Ordovician mass extinction controlled by the rate of climate change. *Science Advances* 11(22): eadv6788.

Zhong Yang-yang, Wu Huai-chun, Fan Jun-xuan, *et al.* 2020. Late Ordovician obliquity-forced glacio-eustasy recorded in the Yangtze Block, South China. *Palaeogeography, Palaeoclimatology, Palaeoecology* 540:109520.

Zhou Yu-ping, Li Yong, Zheng Wang, *et al.* 2024. The role of LIPs in Phanerozoic mass extinctions: An Hg perspective. *Earth-Science Reviews* 249:104667.

Zimmt, J. B., Holland, S. M., Finnegan, S., & Marshall, C. R. 2021. Recognizing pulses of extinction from clusters of last occurrences. *Palaeontology* 64(1):1–20.

Zimmt, J. B., & Jin, J. 2023. A new species of Hirnantia (Orthida, Brachiopoda) and its implications for the Hirnantian age of the Ellis Bay Formation, Anticosti Island, eastern Canada. *Journal of Paleontology* 97(1):47–62.

Zimmt, J. B., Holland, S. M., Desrochers, A., Jones, D. S., & Finnegan, S. 2024. A high-resolution sequence stratigraphic framework for the eastern Ellis Bay Formation, Canada: A record of Hirnantian sea-level change. *Geological Society of America Bulletin* 136(9/10): 3825–3849.

Zou Cai-neng, Qiu Zhen, Poulton, S. W., *et al.* 2018a. Ocean euxinia and climate change "double whammy" drove the Late Ordovician mass extinction. *Geology* 46(6):535–538.

Zou Cai-neng, Qiu Zhen, Wei Heng-ye, Dong Da-zhong, & Lu Bin. 2018b. Euxinia caused the Late Ordovician extinction: Evidence from pyrite morphology and pyritic sulfur isotopic composition in the Yangtze area, South China. *Palaeogeography, Palaeoclimatology, Palaeoecology* 511:1–11.

Acknowledgements

MJM acknowledges ongoing financial support from a Natural Sciences and Engineering Council Discovery Grant. We thank David W. Bapst and an anonymous reviewer for their carefully considered suggestions on an earlier draft, which substantially helped us improve the manuscript.

Cambridge Elements

Elements of Paleontology

Editor-in-Chief

Brenda R. Hunda
Cincinnati Museum Center

About the Series

The Elements of Paleontology series is a publishing collaboration between the Paleontological Society and Cambridge University Press. The series covers the full spectrum of topics in paleontology and paleobiology, and related topics in the Earth and life sciences of interest to students and researchers of paleontology.

The Paleontological Society is an international nonprofit organization devoted exclusively to the science of paleontology: invertebrate and vertebrate paleontology, micropaleontology, and paleobotany. The Society's mission is to advance the study of the fossil record through scientific research, education, and advocacy. Its vision is to be a leading global advocate for understanding life's history and evolution. The Society has several membership categories, including regular, amateur/avocational, student, and retired. Members, representing some 40 countries, include professional paleontologists, academicians, science editors, Earth science teachers, museum specialists, undergraduate and graduate students, postdoctoral scholars, and amateur/avocational paleontologists.

Cambridge Elements

Elements of Paleontology

Elements in the Series

Phylogenetic Comparative Methods: A User's Guide for Paleontologists
Laura C. Soul and David F. Wright

Expanded Sampling Across Ontogeny in Deltasuchus motherali (Neosuchia, Crocodyliformes): Revealing Ecomorphological Niche Partitioning and Appalachian Endemism in Cenomanian Crocodyliforms
Stephanie K. Drumheller, Thomas L. Adams, Hannah Maddox and Christopher R. Noto

Testing Character Evolution Models in Phylogenetic Paleobiology: A Case Study with Cambrian Echinoderms
April Wright, Peter J. Wagner and David F. Wright

The Taphonomy of Echinoids: Skeletal Morphologies, Environmental Factors and Preservation Pathways
James H. Nebelsick and Andrea Mancosu

Virtual Paleontology: Tomographic Techniques For Studying Fossil Echinoderms
Jennifer E. Bauer, Imran A. Rahman

Follow The Fossils: Developing Metrics For Instagram As A Natural Science Communication Tool
Samantha B. Ocon, Lisa Lundgren, Richard T. Bex II, Jennifer E. Bauer, Mary Janes Hughes, and Sadie M. Mills

Niche Evolution and Phylogenetic Community Paleoecology of Late Ordovician Crinoids
Selina R. Cole and David F. Wright

Molecular Paleobiology of the Echinoderm Skeleton
Jeffrey R. Thompson

A Review of Blastozoan Echinoderm Respiratory Structures
Sarah L. Sheffield, Maggie R. Limbeck, Jennifer E. Bauer, Stephen A. Hill, and Martina Nohejlová

A Review and Evaluation of Homology Hypotheses in Echinoderm Paleobiology
Colin D. Sumrall, Sarah L. Sheffield, Jennifer E. Bauer, Jeffrey R. Thompson, and Johnny A. Waters

The Ecology of Biotic Interactions in Echinoids: Modern Insights into Ancient Interactions
Elizabeth Petsios, Lyndsey Farrar, Shamindri Tennakoon, Fatemah Jamal, Roger W. Portell, Michał Kowalewski, and Carrie L. Tyler

What Does Graptolite Origination and Extinction Reveal about the Cause of the Late Ordovician Mass Extinction?
Charles E. Mitchell, H. David Sheets, Michael J. Melchin and Chris Holmden

A full series listing is available at: www.cambridge.org/EPLY

For EU product safety concerns, contact us at Calle de José Abascal, 56–1°,
28003 Madrid, Spain or eugpsr@cambridge.org.

www.ingramcontent.com/pod-product-compliance
Lightning Source LLC
LaVergne TN
LVHW021944060526
838200LV00042B/1918